A 21st Century Employability Skills Improvement Framework for the Construction Industry

This book will provide readers with an understanding of the employability concept and develop an employability skills improvement model to enhance the employability of built environment graduates to foster economic development. The developed model determines the influence of generic skills, discipline-specific skills, work-integrated learning, emotional intelligence, university–industry collaboration outcomes, and 4IR knowledge in predicting the outcomes of improved graduate employability.

The model is developed with a theoretical lens on existing frameworks of employability and skills development. Whilst drawing comparisons with countries such as the United Kingdom, the United States, Australia, and Canada, the authors present the results of a two-stage Delphi survey in South Africa as a case study on the current state of skills development and on the skills of the future. The case study is presented in line with South Africa's long-term National Development Plan (NDP) aimed at developing the key capabilities and skills of its citizens by ensuring quality education on a broader scale by 2030. As automation continues to rapidly advance, the pressures on universities to revamp and restructure their curricula have become increasingly necessary. This book recommends that higher education institutions urgently need to intensify their efforts by introducing significant modifications to the science and technology curriculum to enable students to develop and acquire competencies in the rapidly emerging areas of artificial intelligence, data science, robotics, advanced simulation, data communication, system automation, real-time inventory operations, cloud computing, and information technologies. This implies that universities' curriculum should be infused with 4IR thinking within the conventional primary sciences of biology, chemistry, and physics, with greater emphasis on digital literacy to boost 4IR understanding amongst the graduates.

The book is therefore of interest to researchers and policymakers in the built environment that are placed in academia, the construction industry, or at consultancy levels; it provides significant recommendations for universities as they intensify their efforts to develop graduates for the future.

John Aliu is a vibrant and industrious civil engineer by profession. He obtained his BEng in Civil Engineering in 2010 from Nnamdi Azikiwe University, Nigeria. Thereafter, he worked as a civil engineer between 2013 and 2016. He obtained his Master's Degree in Construction Management in 2017 from the University of Johannesburg, South Africa. Finishing with a Distinction during his master's earned him a Golden Key International Honour membership. He then obtained his PhD in Civil Engineering also from the University of Johannesburg in 2020. Dr. Aliu is currently a Postdoctoral Fellow at the Faculty of Engineering and the Built Environment at the CIDB Centre of Excellence, University of Johannesburg, South Africa. He shares research interest in construction education, engineering education, skills development, project management, and learning in the era of the fourth industrial revolution. He also serves as a regular reviewer for various ASCE, Elsevier, Emerald, Springer and Taylor & Francis built environment journals. He is a member of Project Management South Africa (PMSA) and South African Institution of Civil Engineering (SAICE).

Clinton Aigbavboa is a Professor at the Department of Construction Management and Quantity Surveying, University of Johannesburg, South Africa. Before joining academia, he was involved as a quantity surveyor on several infrastructural projects, both in Nigeria and in South Africa. Prof. Aigbavboa is the immediate past Vice Dean of the Faculty of Engineering and Built Environment, University of Johannesburg, South Africa. He has extensive knowledge in practice, research, training, and teaching. He is currently the Director of the Construction Industry Development Board Centre of Excellence and the Sustainable Human Settlement and Construction Research Centre at the University of Johannesburg. He is also an author of seven research books that were published with Springer Nature and CRC Press. He is Editor-in-Chief of the *Journal of Construction Project Management and Innovation* (accredited by the DoHET) and has received national and international recognition in his field of research. He is rated by the South Africa National Research Foundation.

Wellington Thwala is a Professor at the Department of Construction Management and Quantity Surveying, University of Johannesburg, South Africa. He is the immediate past Head of the Department of Construction Management and Quantity Surveying, University of Johannesburg, South Africa. Currently, he is the Chair of SARChI in Sustainable Construction Management and Leadership in the Built Environment, FEBE, University of Johannesburg, South Africa. He offers research support and advice on construction-related issues to the Construction Industry in South Africa and government. Prof. Thwala has extensive experience of providing consultancy for project leadership and management of construction projects and teaching project management subjects at the postgraduate level. He has an extensive industry experience with research focus on sustainable construction, leadership, and project management. He is Editor-in-Chief of the *Journal of Construction Project Management and Innovation*. He also serves as an editorial board member to various reputable international journals.

Routledge Research Collections for Construction in Developing Countries

Series Editors: Clinton Aigbavboa, Wellington Thwala, Chimay Anumba, David Edwards

A 21st Century Employability Skills Improvement Framework for the Construction Industry

John Aliu, Clinton Aigbavboa and Wellington Thwala

A 21st Century Employability Skills Improvement Framework for the Construction Industry

John Aliu
Clinton Aigbavboa
Wellington Thwala

R Routledge
Taylor & Francis Group

LONDON AND NEW YORK

First published 2021
by Routledge
2 Park Square, Milton Park, Abingdon, Oxon OX14 4RN

and by Routledge
52 Vanderbilt Avenue, New York, NY 10017

Routledge is an imprint of the Taylor & Francis Group, an informa business

British Library Cataloguing-in-Publication Data
A catalogue record for this book is available from the British Library

Library of Congress Cataloging-in-Publication Data
A catalog record for this book has been requested

ISBN: 978-0-367-68401-3 (hbk)
ISBN: 978-0-367-68430-3 (pbk)
ISBN: 978-1-003-13750-4 (ebk)

Typeset in Goudy
by Apex CoVantage, LLC

Contents

Figures

Tables

Preface

The present-day construction industry has experienced dramatic changes from what it was some decades ago. Factors such as increased global competition, the sophistication of construction projects, and a rise in technology development have rendered the sector more complex. Most recently, the fourth industrial revolution (4IR) has placed significant pressures on higher education to produce adequately skilled and knowledgeable graduates to handle the ever industry-increasing expectations. Hence, universities face a daunting task to equip students with more than just subject knowledge but also with work-readiness skills to seamlessly transit into the world of work. This book develops a 21st Century Employability Skills Improvement Framework for the construction industry, thereby fostering economic development. The framework is developed to ascertain the influence of generic skills, discipline-specific skills, work-integrated learning, emotional intelligence, university–industry collaboration outcomes, and 4IR knowledge in predicting the outcomes of improved graduate employability. The framework is developed on the basis of the theory developed from the review of literature and a two-stage Delphi survey. Subsequently, a questionnaire survey was conducted to validate the conceptual framework. The study was conducted in the Gauteng Province of South Africa and is in line with the nation's long-term National Development Plan (NDP) that aims to develop the key capabilities and skills of its citizenry by ensuring quality education on a broader scale by 2030.

For the Delphi study, 14 experts completed a two-stage process. Using a five-factor criterion, these experts from both academia and industry identified 230 measured variables. Data obtained through the Delphi process assisted in validating the various dimensions (main and sub-dimensions) that contribute to the employability of built environment graduates, which ultimately results in the employability framework. The Delphi method also determined the outcomes of improved graduate employability in meeting the expectations of the construction industry. The Delphi method was considered the best method to address these objectives which aimed at soliciting expert opinions on factors that contribute to the development of an employability framework for built environment graduates. Apart from the Delphi study, focus groups were also considered. However, focus group discussions were not adopted because getting the selected experts to converge in one place at a particular time to deliberate was cumbersome, costly, and not feasible. Moreover, the adoption of the Delphi method eliminates bias as the members of the expert panel remain anonymous to each other and cannot be influenced by other members of the panel. This is not possible when adopting the focus group approach. The expected output is the extent to which graduate employability is influenced by certain established factors, and consensus on the critical factors that contribute to improved graduate employability of built environment graduates using South Africa as a case study.

The first round questionnaire was designed on the basis of the literature review to determine the various dimensions (main and sub-dimensions) that contribute to the employability of built environment graduates in South Africa. The experts' responses and opinions were presented and analysed on Microsoft Excel spreadsheet and the median, mean, standard deviation, and the interquartile values were observed. In the second round, a questionnaire was formulated on the basis of the responses and opinions from the first round. The second round presented an opportunity for experts to review, rank, and make comments on the various dimensions (main and sub-dimensions) that contribute to the employability of built environment graduates which were addressed by the experts in the first round. Moreover, the questions in the second round were close-ended which provided the opportunity for experts to agree, disagree, or seek clarifications concerning the outlined dimensions that contribute to improved graduate employability. Based on recommendations by some experts, identical factors were reworded. This was achieved using a ten-point Likert scale of 'no significance', 'low significance', 'medium significance', 'high significance', and 'very high significance', while 'no significance' was assigned the lowest weighting (1 and 2) and 'very high significance' had the highest weighting (9 and 10). The degrees of consensus reached amongst experts regarding the outlined dimensions that contribute to improved graduate employability were measured by the frequencies obtained. Based on analyses of the second round, the dimensions that contribute to improved graduate employability were prepared, which conceptualizes the framework for the broader study. For every round, the respective group median value and interquartile deviation (IQD) were computed as a measure of the central tendencies to determine consensus. Owing to its tendency to minimize the influence of potentially biased persons, the median value was adopted as a measure of central tendency as opposed to the mean and IQD. On the other hand, the respective IQD scores helped to summarize data variability and hence identified the appropriate measures that influenced graduate employability. By eliminating outlying values, the IQD also offers general coherence and clarity of the overall dataset. The interquartile range, which is closely related to the median, indicates the extent to which the central 50% of values within the presented dataset are dispersed. After this second round, consensus was reached.

This book made a significant contribution to the body of knowledge. Theoretically, this book developed a priori employability framework using six latent constructs with the inclusion of two new constructs: university–industry collaboration outcomes and 4IR knowledge. It is worthwhile to note that previous employability models have used other constructs without the inclusion of those adopted for this study and in various contexts. From the developed employability framework, it was possible to ascertain which of the latent constructs contribute and predict graduate employability, something which most existing studies have struggled to achieve. Therefore, this study proved that apart from attaining an academic degree, several other factors influence graduate employability. Hence, this book provides a foundation for other researchers who will conduct future employability studies. As automation continues to rapidly advance, the pressures on universities to revamp and restructure its curriculum (skills revolution) have become increasingly necessary. This research book recommends that higher education institutions urgently need to intensify their efforts by introducing significant modifications to the science and technology curriculum to enable students to develop and acquire competencies in the rapidly emerging areas of artificial intelligence, data science, robotics, advanced simulation, data communication, system automation, real-time inventory operations, cloud computing, and information technologies. This implies that universities' curriculum should be infused with 4IR thinking within the conventional primary sciences of biology, chemistry, and physics, with greater emphasis on digital literacy to boost 4IR understanding amongst the graduates. The book is therefore of interest to researchers and policymakers in the built environment that are placed in academia, construction industry, or at consultancy levels. Furthermore, the authors confirm that the text utilized in this work reflects original work and, where necessary, the material has benefited from relevant context setting and referencing.

Acknowledgements

We express our appreciation and befittingly acknowledge all those who made significant contribution to this book from the conceptualization stage to the completion stage. Special mention must be made to the Department of Construction Management and Quantity Surveying and the Civil Engineering Department, University of Johannesburg, South Africa, for providing the opportunity to successfully embark on this study. Finally, we also express our gratitude to the CIDB Centre of Excellence for the Postdoctoral Fellowship granted which paved way for the contribution of this book.

Abbreviations

4IR – Fourth industrial revolution
ABET – Accreditation Board for Engineering and Technology
ABET – Adult Basic Education and Training
ACCI – Australian Chamber of Commerce and Industry
AI – Artificial intelligence
AMP – Advanced manufacturing partnership
AMT – Advanced manufacturing technologies
ANC – African National Congress
ASGISA – South Africa Accelerated and Shared Growth Initiative
ASTD – American Society for Training and Development
BCA – Business Council of Australia
BIM – Building information modelling
CBE – Council of the Built Environment
CDL – Career Development Learning
CEM – Council of Education Ministers
CHE – Council on Higher Education
CHET – Centre for Higher Education Transformation
CME – Canadian Manufacturers and Exporters
CPS – Cyber-physical systems
CSR – Corporate social responsibility
CTP – Committee of Technikon Principals
CUP – Committee of University Principals
DBE – Department of Basic Education
DeSeCo – Definition and Selection of Competencies
DfES – Department for Education and Skills
DHET – Department of Higher Education and Training
DIPP – Department of Industrial Policy and Promotion
DoE – Department of Education
DSO – Delphi-specific objectives
DSS – Discipline-specific skills
DTI – Department of Trade and Industry
ECSA – Engineering Council of South Africa
EFFRA – European Factories of the Future Research Association

EI – Emotional intelligence
EIF – Employability Improvement Framework
EPU – Economic Planning Unit
ETQA – Education Training Quality Assurer
FETC – Further Education and Training Certificate
FoF – Factories of the Future
GCSE – General Certificate of Secondary Education
GET – General Education and Training
GS – Generic skills
HC – Human capital
HCT – Human Capital Theory
HEDCOM – Heads of Education Departments Committee
HEIs – Higher education institutions
HEQC – Higher Education Quality Committee
HEQF – Higher Education Qualification Framework
HRD – Human resource development
HSRC – Human Science Research Council
IT – Industrial training
ICT – Information communication and technology
IGE – Improved graduate employability
IIoT – Industrial Internet of Things
IoS – Internet of Services
IoT – Internet of Things
IP – Internet Protocol
IPAP – Industrial Policy Action Plan
IQD – Interquartile range
IT – Institutional theory
JIPSA – Joint Initiative for Priority Skills Acquisition
LT – Learning theory
MAE – Mean absolute error
MINETUR – Ministry of Industry, Energy and Tourism
MIUR – Ministry of Education, Universities and Research
MMA – Mixed-method approach
MOTIE – Ministry of Trade, Industry and Energy
MTEF – Medium-Term Expenditure Framework
NCHE – National Commission on Higher Education
NCVER – National Centre for Vocational Education Research
NDP – National Development Plan
NEPA – National Education Policy Act
NGP – New Growth Path
NQF – National Qualification Framework
NRF – National Research Foundation
NSB – National Standards Bodies
NSDS II – National Skills Development Strategy II
NSF – National Science Foundation

NTB – National Training Board
NZCF – New Zealand Curriculum Framework
OBE – Outcome-based education
OECD – Organisation for Economic Co-operation and Development
PBL – Project-based learning
PCAST – President's Council of Advisors on Science and Technology
PLC – Programmable logic controllers
PMSA – Project and construction management professions
RD – Resource dependency
RFID – Radio-frequency identification
RPL – Recognition of Prior Learning
SACE – South African Council for Educators
SACPCMP – South African Council for the Project and Construction Management Professions
SAICE – South African Institution of Civil Engineering
SAQA – South African Qualification Framework
SASA – South African Schools Act
SAUVCA – South African Universities Vice Chancellors Association
SC – Strategic choice
SCANS – Secretary's Commission on Achieving Necessary Skills
SEM – Structural equation modelling
SGB – Standards Generating Bodies
ST – Stakeholder theory
STCPDC – School-to-Careers Professional Development Centre
TAFE – Technical and Further Education
TCE – Transaction cost economics
TH – Triple helix
TVET – Technical and Vocational Education and Training
UK – The United Kingdom
US – The United States
UIC – University–industry collaboration
UNESCO – The United Nations Educational, Scientific and Cultural Organization
UNISA – University of South Africa
VET – Vocational education and training
WEF – World Economic Forum
WIL – Work-integrated learning

Section I
Background information

1 General introduction

Introduction

The construction industry is instrumental in developing the built environment of economies around the world, and the South African construction industry is not exempted from this. Its numerous activities, including the construction and maintenance of highways, reservoirs, bridges, and structures, amongst others, are crucial in achieving the various socio-economic needs of countries worldwide. In the same vein, it is the sector of the economy which transforms resources into constructed facilities through planning, design, construction, maintenance, and sustainability. However, the present-day construction industry has experienced dramatic changes from what it was some decades ago. Factors such as increased global competition, sophistication, complexities of construction projects, and a rise in technology development constantly render the sector more complex (Tatum, 2010; Aliu and Aigbavboa, 2017). Other factors include increased project targets and deliveries, changing demographics, environmental impact, shrinking product life cycles, and changing organizational structures and designs, amongst others. With the global population continually skyrocketing and expected to reach around 9.7 billion by 2050 (UNESCO, 2018), various pressures such as global warming, infrastructural development, energy use, and sustainable practices, amongst other factors, will significantly increase (Baller *et al.*, 2016). Hence, the dynamism of today's workplace makes the sector heavily reliant on higher education institutions (HEIs) (universities) and training programmes to produce adequately skilled and knowledgeable graduates to handle their increasing expectations. This increases the need for universities to equip students with more than just subject knowledge but also work-readiness skills to enable them to transit seamlessly into the world of work. According to Hyland (2019), both the education and training sectors are presented with the daunting task of aligning themselves with the construction industry to be conversant with the ever-changing skill demands of employers. In modern times, the functions of higher education and training sectors have been increasingly and successfully linked to national productivity and economic growth (Aliu and Aigbavboa, 2017). This implies that universities globally must continuously deliver high-quality construction pedagogy to produce graduates who are skilled and performance orientated. More than ever, universities have been identified as a vehicle for the advancement of the national economy and society as a whole. Their roles have seen them dubbed as 'knowledge creators' as they are key in preparing graduates with appropriate industry skills and knowledge. Hence, this book proposes that by furnishing students with the required employability skills in universities, they are better prepared and equipped to take up industry positions after graduation. Moreover, there will also be a greater pool of potentially skilled graduates from which industry employers can choose.

As stated earlier, the various complexities of the construction industry require present-day graduates to be significantly prepared to integrate academic knowledge acquired in lecture rooms to deal with arising industry problems. They must be able to employ holistic approaches in problem-solving scenarios by utilizing creative and critical thinking skills. In an increasingly global workplace, graduates will be expected to have a global perspective and be ready to work with various teams in achieving timely results (Archer and Davison, 2008). They must also possess good communication skills to interact clearly and confidently within the professional confines (Washer, 2007; Archer and Davison, 2008; Arain, 2010). Today's ethical issues are assuming global proportions, and graduates must possess a strong ethical foundation to deal with issues involving the equitable distribution of resources, by-products of design, sustainable development, and environmental conservation, amongst others (Arain, 2010; Mat and Zabidi, 2010; Ma and Sun, 2013). They also need to be conversant with the legal aspects of engineering and its social impact as well as have a proper understanding of business practices and entrepreneurship ideas (Jackson and Chapman, 2012). The possession of these skills and knowledge will enhance their employment prospects as well as the understanding of the socio-economic implications of their chosen careers. Given that present-day industry employers constantly seek graduates with industry skills and knowledge, nations that invest in quality construction education place their citizens at a distinct advantage. Economically, the possession of requisite industry knowledge increases efficiency, competitiveness, and productivity of the industry workforce (Rychen and Salganik, 2000; Aliu and Aigbavboa, 2017). This was amplified in a report published by the World Economic Forum held in Cape Town, South Africa, in 2019, which recommends that nations should strive to ensure that their educational systems produce competent graduates who can sustain the future of the construction industry. This implies that the prosperity of any nation depends on the contributions of its education and training sector. Simply put, a generally informed nation will be a productive one. Likewise, the possession of both academic and non-academic skills is pivotal in preparing students to perform at the top level in their various careers after graduation (Aliu and Aigbavboa, 2017). This argument firmly relates to the concept of the human capital theory, which argues that the better skilled a graduate is at a workplace, the higher the level of efficiency and effectiveness in performing industry tasks. This infers that the future of any nation depends on the specialized skills, abilities, and competencies of its workforce in proffering solutions to arising industry problems. Hence, this book is aimed at developing an employability improvement framework for built environment graduates to meet the increasing and ever-changing needs of the construction industry. Consequently, this book is relevant because higher education institutions around the world are expected to foster learning outcomes that are valued by the employers of the construction industry. This book also points to the fact that adequately skilled graduates will profit the construction industry and society as well as sustain the built environment as the world experiences a paradigm shift towards the fourth industrial revolution (4IR). Most recently, the 4IR has presented a new dimension to the construction industry, paving the way for even more complexities and hence placing significant pressures on higher education.

The idea of 'employability' is a broad concept that implies an extensive diversity of proficiencies needed by a student to effectively function in the construction industry. McQuaid and Lindsay (2005) indicate that employability relates to both the employed and the unemployed. In recent years, nations around the world have experienced an upsurge in the flexibility of the labour market, a scenario that has contributed immensely to the change in the demand and supply of conditions for employment. This structural adjustment has paved the way for new careers and positions within the industry, which requires job seekers to possess an improved level of knowledge, skills, and competencies (Danson, 2005; Räty *et al.*, 2019). In this case, this book postulates that built environment students will be confronted by workplace dynamics and challenges, which increases the need for them to be adequately prepared to handle industry challenges and responsibilities after graduation. This book, therefore, argues that when a graduate possesses a wide variety of knowledge, skills, and competencies relevant to the 21st century in the changing context of the 4IR, they will become more employable and relevant in a changing and challenging work environment. It is worth noting that the subject of employability has been discussed across several contexts in higher education (Fallows and Steven, 2000; Mason *et al.*, 2006; Dacre Pool and Sewell, 2007; Baker and Henson, 2010; Hinchliffe and Jolly, 2011; Torres-Machi *et al.*, 2013; Oliver, 2015; Pitan, 2016; Aliu and Aigbavboa, 2017; Donald *et al.*, 2019). Its role in the context of higher education has also become a cause for global concern in recent times (Jonck, 2014). The concept has been explored and investigated across several developing and developed countries such as Malaysia (Omar *et al.*, 2012; Hanapi and Nordin, 2014), Australia (Nagarajan and Edwards, 2015), Namibia (Naanda, 2010), Taiwan (Pan and Lee, 2011), India (Mishra, 2014), Botswana (Alao *et al.*, 2009), Australia (Curtis and McKenzie, 2001), the United States (Holdsworth and Gearhart, 2002), Germany (Gibbons-Wood and Lange, 2000), the United Kingdom (Saunders and Zuzel, 2010; Benson *et al.*, 2014), Nigeria (Pitan, 2016), and South Africa (Griesel and Parker, 2009; Bezuidenhout, 2011; Symington, 2012; Kundaeli, 2016).

As noted earlier, the present-day construction industry can be described as fast-paced, dynamic, flexible, and ever-changing, which places optimum pressures on graduates to function effectively in various capacities in which they find themselves (Fallows and Steven, 2000; Fugate and Kinicki, 2008). It has been acknowledged that this era of work has placed various difficulties on both employers of labour and its employees who were previously absent, or has increased the pressures that are currently being experienced. The construction industry employers now seek graduates who not only are academically sound but also possess non-academic skills, knowledge, and competencies. This indicates that the availability of skilled graduates is extremely important to present-day industry employers. This high expectation also reduces the promise of job security for employees. Job-seeking graduates, on the other hand, can no longer depend on their academic certificates to obtain dream jobs or employment opportunities (Knight, 2006). Owing to the global economic situation, the need for skilled and knowledgeable graduates is rapidly increasing. It is therefore essential that both the graduate and the unemployed population are conscious of the significant employability skills and competencies that are now speedily becoming the principal indicators in the skilled and semi-skilled discourse. Subsequently, industry employers have continually expressed their discontent regarding the disparity between the qualities of the educational outcomes of graduates and the increasing needs of both the workplace and society. The assumption is that present-day graduates lack the appropriate employability skills to function effectively in the construction industry after graduation. This has prompted employers to re-train graduates to provide them with the necessary skills to thrive and improve their existing job competency levels. Again, there are some disagreements regarding employability skills for graduates, which may be attributed to the fluctuating industry requirements across the world, which has been largely triggered by the recent wave of digitalization. Nonetheless, as significant as the variances may be, there is a unanimous perception that graduates must possess certain employability skills to remain relevant in the construction industry.

Aims and objectives of the book

Considering the present-day dynamics facing the construction industry, HEIs across the world are under significant pressure to respond even more effectively to the rapidly changing landscape. Despite the efforts of several studies to discuss the issue of graduate employability, the development of an employability improvement framework has not been considered in the built environment context in light of the 4IR movement. This study focused on built environment graduates who are poised to take up industry positions after graduation. Furthermore, the study examined various pedagogical approaches in developing employability skills and predicted the undergraduate courses that will be relevant to the built environment in the next ten years owing to the advent of the 4IR. Ultimately, the study determined the influence of several latent constructs such as generic skills, discipline-specific skills, work-integrated learning, emotional intelligence, university–industry collaboration outcomes, and 4IR knowledge on graduate employability. As no employability framework for built environment graduates has been developed, this research book filled the gap. The various dimensions (exogenous variables) of the proposed framework include generic skills, discipline-specific skills, work-integrated learning, emotional intelligence

(EI), university–industry collaboration, and Industry 4.0 requirements, the latter two being new dimensions considered by the proposed framework and are absent in existing models across any context. It is worth noting that previous employ-ability models have not tested these latent constructs in a multidimensional structure. As an ultimate objective, this study determined the possible outcomes of improved graduate employability in meeting the expectations of the construc-tion industry.

Significance of the book

This book will provide readers with an understanding of the employability con-cept and likewise develop an Employability Skills Improvement Framework to enhance the quality of built environment graduates to foster economic devel-opment. The developed framework ascertains the influence of generic skills, discipline specific skills, work-integrated learning, emotional intelligence, uni-versity–industry collaboration outcomes, and 4IR knowledge in predicting the outcomes of improved graduate employability. The framework is developed with a theoretical lens on existing frameworks of employability and skills develop-ment. From the reviewed literature, it appears that an employability improve-ment framework for built environment graduates does not exist. This book fills this gap from a South African perspective. The South African educational sector has a vital role to play in realizing the nation's future goals. It aims at develop-ing 'one million young people in various Industry 4.0 elements and related skills by 2030', according to the nation's President Cyril Ramaphosa. In achieving this milestone, the nation's commitment to improving the curriculum of higher education is certain, with various proposed initiatives put in place by the gov-ernment. What is significant though is the nation's goal for its graduates to be adequately skilled to handle industry positions owing to the advent of the 4IR. This is in line with the nation's long-term National Development Plan (NDP) that aims to develop the key capabilities and skills of its citizenry by ensuring quality education on a broader scale by 2030. More so, the quality education and skills development mandate of the NDP aligns with the United Nations' Sustain-able Development Goals (SDGs), which were adopted in September 2015 by the General Assembly. One of the UN's resolutions is to ensure quality and inclusive education by enabling effective learning and conducive environments to encour-age lifelong learning for all.

This book therefore contributes to the body of employability knowledge by introducing the 4IR component into the employability discourse. This book determines the skills that will be most sought after by employers in light of the 4IR, as well as predicts the undergraduate courses that will become relevant to the built environment in the next ten years. This book is also significant owing to the adoption of structural equation modelling (SEM) using the partial least squares (PLS) approach in analysing and developing an employability improve-ment framework. Owing to the advent of the 4IR, there have been several gener-alized statements as to which of the latent constructs contribute to graduate employability. Therefore, by conducting this study using PLS-SEM, it was possible to identify precisely the constructs that are statistically significant and contribute to graduate employability such as generic skills, discipline-specific skills, work-integrated learning, emotional intelligence, university–industry collaboration outcomes, and 4IR knowledge.

This employability framework is timely, given the increased need for nations globally to develop a competent future workforce to handle industry positions owing to the advent of the 4IR. The findings of this research book are influential to the educational sector in both the developing and developed countries, most especially in the formulation and introduction of 4IR courses (modules), guidelines, and philosophies for both private and public centres of higher learning. Findings from the book will provide construction educators and academicians with the opportunities to make knowledgeable decisions on the built environment curricula so that developing skills becomes an absolute necessity. The additional knowledge derived from this study will be beneficial in achieving a more realistic understanding of the factors that affect the employability of built environment graduates in South Africa and beyond. For individuals who are involved in skilled human resource development, this book is valuable because it enlightens policymakers on employability skills that graduates require to fit easily into the world of work. The findings emanated from this book reinforce the need to establish effective collaborations between higher education and the construction industry to produce competent graduates who can effectively thrive after graduation. More importantly, the impact of this research will be better appreciated if appropriately executed by the necessary stakeholders. As automation continues to rapidly advance, the pressures on universities to revamp and restructure its curriculum (skills revolution) have become increasingly necessary. This research book recommends that higher education institutions urgently need to intensify their efforts by introducing significant modifications to the science and technology curriculum to enable students to develop and acquire competencies in the rapidly emerging areas of artificial intelligence, data science, robotics, advanced simulation, data communication, system automation, real-time inventory operations, cloud computing, and information technologies. This implies that universities' curriculum should be infused with 4IR thinking within the conventional primary sciences of biology, chemistry, and physics, with greater emphasis on digital literacy to boost 4IR understanding amongst the graduates. The book is therefore of interest to researchers and policymakers in the built environment that are placed in academia, construction industry, or at consultancy levels.

Structure of the book

For guidance and ease of use, this book is divided into nine chapters. This first chapter consists of detailing the background information for the book by examining critical sections such as the aims and objectives, significance and relevance of the book. Chapter 2 presents the theoretical and conceptual frameworks (perspectives) underpinning the study by conducting a rigorous literature review. This chapter also reviews the history of employability as well as relevant models, which provided key variables in developing the employability framework. While Chapter 3 extensively addresses the gaps in employability studies, Chapter 4 reviews the concept of learning. Furthermore, Chapter 5 reviews the various employability and generic competency frameworks in several international contexts and Chapter 6 provides an overview of the educational system in South Africa. Finally, Chapter 7 discusses the conceptual framework, Chapter 8 validates the constructs while Chapter 9 provides conclusions and recommendations.

Summary

This chapter introduces the idea behind the conception of this research book with emphasis on issues surrounding the employability concept in both the generic sense and the fourth industrial revolution (4IR) context. A conscious attempt was made to explain the relevance and timeliness of this book. The main objective of this book is to develop an Employability Skills Improvement Framework for the construction industry. The framework will ascertain the influence of generic skills, discipline-specific skills, work-integrated learning, emotional intelligence, university–industry collaboration outcomes, and 4IR knowledge in predicting the outcomes of improved graduate employability. This book addresses employability from the built environment perspective and only refers to other disciplines during the generic literature review. The next chapter presents the theoretical and conceptual employability frameworks by conducting a rigorous literature review, which provides key variables in developing the employability framework.

References

Alao, A.A., Pilane, C.D., Mabote, M.M., Setlhare, K., Mophuting, K., Semphadile, K.M., Oridile, L.W., Kgathi, P.L. and Mmapatsi, S. 2009. Employer satisfaction survey of the University of Botswana graduates. *International Journal of Contemporary Issues in Education and Psychology*, 1(3).

Aliu, O.J. and Aigbavboa, C.O. 2017. *Upscaling construction education to meet the needs of the Nigerian construction industry* (Master's dissertation, University of Johannesburg). Retrieved from: ujcontent.uj.ac.za

Arain, F.M. 2010. Identifying competencies for baccalaureate level construction education: Enhancing employability of young professionals in the construction industry. In *Proceedings of the Construction Research Congress 2010: Innovation for Reshaping Construction Practice*. Construction Research Congress 2010, May 8–10, 2010, American Society of Civil Engineers (ASCE), Banff, AB, pp. 194–204.

Archer, W. and Davison, J. 2008. *Graduate employability: What employers want?* London: The Council for Industry and Higher Education.

Baker, G. and Henson, D. 2010. Promoting employability skills development in a research-intensive university. *Education + Training*, 52(1), pp. 62–75.

Baller, S., Dutta, S. and Lanvin, B. 2016. *Global information technology report 2016*. Geneva, Switzerland: Ouranos.

Benson, V., Morgan, S. and Filippaios, F. 2014. Social career management: Social media and employability skills gap. *Computers in Human Behavior, 30*, pp. 519–525.

Bezuidenhout, M. 2011. *The development and evaluation of a measure of graduate employability in the context of the new world of work* (Doctoral dissertation, University of Pretoria, Pretoria). Retrieved from: http://hdl.handle.net/2263/28552

Bhanugopan, R. and Fish, A. 2009. Achieving graduate employability through consensus in the South Pacific island nation. *Education Training, 51*(2), pp. 108–123. http://dx.doi.org/10.1108/00400910910941273.

Curtis, D. and McKenzie, P. 2001. *Employability skills for Australian industry: Literature review and framework development*. Melbourne: Australian Council for Educational Research.

Dacre Pool, L. and Sewell, P. 2007. The key to employability: Developing a practical model of graduate employability. *Education Training, 49*(4), pp. 277–289. https://doi.org/10.1108/00400910710754435.

Danson, M. 2005. Old industrial regions and employability. *Urban Studies, 42*(2), pp. 285–300.

Donald, W.E., Baruch, Y. and Ashleigh, M. 2019. The undergraduate self-perception of employability: Human capital, careers advice, and career ownership. *Studies in Higher Education, 44*(4), pp. 599–614.

Fallows, S. and Steven, C. 2000. Building employability skills into the higher education curriculum: A university-wide initiative. *Education Training, 42*(2), pp. 75–83. http://dx.doi.org/10.1108/00400910010331620.

Fugate, M. and Kinicki, A.J. 2008. A dispositional approach to employability: Development of a measure and test of implications for employee reactions to organizational change. *Journal of Occupational and Organizational Psychology, 81*, pp. 503–527.

Gibbons-Wood, D. and Lange, T. 2000. Developing core skills: Lessons from Germany and Sweden. *Education and Training, 42*(1), pp. 24–32.

Griesel, H. and Parker, B. 2009. Graduate attributes. *Higher Education South Africa and the South African Qualifications Authority*. Pretoria.

Hanapi, Z. and Nordin, M.S. 2014. Unemployment among Malaysia graduates: Graduates' attributes, lecturers' competency and quality of education. *Procedia: Social and Behavioural Sciences, 112*, pp. 1056–1063.

Hinchliffe, G.W. and Jolly, A. 2011. Graduate identity and employability. *British Educational Research Journal, 37*(4), pp. 563–584.

Holdsworth, T. and Gearhart, E. 2002. Teaching and assessing employability skills. *Modern Machine Shop, 74*(12), p. 158.

Hyland, T. 2019. *Vocational studies, lifelong learning and social values: Investigating education, training and NVQs under the new deal*. Abingdon, UK: Routledge.

Jackson, D. and Chapman, E. 2012. Non-technical competencies in undergraduate business degree programs: Australian and UK perspectives. *Studies in Higher Education, 37*(5), pp. 541–567. doi: 10.1080/03075079.2010.527935.

Jonck, P. 2014. A human capital evaluation of graduates from the Faculty of Management Sciences employability skills in South Africa. *Academic Journal of Interdisciplinary Studies, 3*(6), pp. 265–274. https://doi.org/10.5901/ajis.2014.v3n6p265.

Knight, P. 2006. *Embedding employability into the curriculum*. Learning and employability series, no. 1. York: Higher Education Academy. Retrieved from: http://www.heacademy.ac.uk/assets/documents/employability/id460_embedding_employability_into_the_curriculum_338.pdf

Kundaeli, F. 2016. *Individual factors affecting the employability of information systems graduates in Cape Town, South Africa: Employed graduates and employer perspectives* (Doctoral dissertation, University of Cape Town, Cape Town). Retrieved from: http://hdl.handle.net/11427/22891

Ma, X. and Sun, Y. 2013. Research on employment ability evaluation of graduates. *Proceedings of the 19th International Conference on Industrial Engineering and Engineering Management*, Springer, p. 1019.

Mason, G., Williams, G. and Cranmer, S. 2006. Employability skills initiatives in higher education: What effects do they have on graduate labour market outcomes? *Education Economics*, 17(1), pp. 1–30.

Mat, N.H.N. and Zabidi, Z.N. 2010. Professionalism in practices: A preliminary study on Malaysian public universities. *International Journal of Business and Management*, 5(8), p. 138.

McQuaid, R.W. and Lindsay, C. 2005. The concept of employability. *Urban Studies*, 42(2), pp. 197–219.

Mishra, K. 2014. Employability skills that recruiters demand. *IUP Journal of Soft Skills*, 8(3), p. 50.

Naanda, R.N. 2010. *The integration of identified employability skills into the Namibian vocational education and training curriculum* (Doctoral dissertation, University of Stellenbosch, Stellenbosch).

Nagarajan, S. and Edwards, J. 2015. The role of universities, employers, graduates and professional associations in the development of professional skills of new graduates. *Journal of Perspectives in Applied Academic Practice*, 3(2).

Oliver, B. 2015. Redefining graduate employability and work-integrated learning: Proposals for effective higher education in disrupted economies. *Journal of Teaching and Learning for Graduate Employability*, 6(1), pp. 56–65.

Omar, N.H., Abdul Manaf, A., Helma Mohd, R., Che Kassim, A. and Abd Aziz, K. 2012. Graduates' employability skills based on current job demand through electronic advertisement. *Asian Social Science*, 8(9), pp. 103–110.

Pan, Y.J. and Lee, L.S. 2011. Academic performance and perceived employability of graduate students in business and management: An analysis of nationwide graduate destination survey. *Procedia-Social and Behavioural Sciences*, 25, pp. 91–103.

Pitan, O.S. 2016. Towards enhancing university graduate employability in Nigeria. *Journal of Sociology and Social Anthropology*, 7(1), pp. 1–11.

Räty, H., Kozlinska, I., Kasanen, K., Siivonen, P., Komulainen, K. and Hytti, U. 2019. Being stable and getting along with others: Perceived ability expectations and employability among Finnish university students. *Social Psychology of Education*, pp. 1–17.

Rychen, D.S. and Salganik, L.H. 2000. Definition and selection of key competencies. *The Fourth General Assembly of the OECD Education Indicators Programme, the INES Compendium, Contributions from the INES Networks and Working Groups*. Paris: The Organisation for Economic Co-operation and Development, pp. 61–73.

Saunders, V. and Zuzel, K. 2010. Evaluating employability skills: Employer and student perceptions. *Bioscience Education*, 15(1), pp. 1–15.

Symington, N. 2012. *Investigating graduate employability and psychological career resources* (Doctoral dissertation, University of Pretoria, Pretoria).

Tatum, C. 2010. Construction engineering education: Need, content, learning approaches. In *Proceedings of the Construction Research Congress 2010: Innovation for Reshaping Construction Practice*, Construction Research Congress 2010, May 8–10, 2010. American Society of Civil Engineers (ASCE), Banff, AB, p. 183.

Torres-Machi, C., Dahan, A., Yepes, V. and Pellicer, E. 2013. Comparative study of employability between Spanish and French students in civil engineering. *Journal of Professional Issues in Engineering Education and Practice* (American Society of Civil Engineers, ISSN-1052-3928), *139*(2), pp. 163–170.

UNESCO, U. 2018. *Global education monitoring report.* Global education monitoring report gender review: Meeting our commitments to gender equality in education.

Washer, P. 2007. Revisiting key skills: A practical framework for higher education. *Quality in Higher Education, 13*(1), pp. 57–67. doi: 10.1080/13538320701272755.

Section II

Employability theories and model development

2 Theoretical background surrounding employability

Theoretical framework

Human capital theory

As a result of the rapidly developing and ever-changing nature of the construction industry, there is an urgent requirement for graduates who are adequately skilled. Globally, it has been emphasized that it is the sole duty of universities to furnish students with the required skills, knowledge, and competencies. This increases the pressure on the academic setting and its educators with regard to the quality of learning and teaching at the higher education level. As a result of such unprecedented pressure on universities to deliver on their mandate of producing quality graduates, the emphasis on educational approaches has evolved. This is because the required skills students need to develop to solve industry problems effectively today are different from what was needed decades ago. Present-day industry employers seek graduates with good academic skills in addition to numerous non-academic skills, including communication, problem-solving, interpersonal team skills, and leadership skills, amongst others. Hence, learning design approaches, as well as conducive learning environments, must support the development of the skills mentioned previously. Thus, the urgency to find innovative and pedagogical ways of developing present-day students to ensure a seamless transition into industry positions has taken on a new dimension. As a result, several modern educational strategies such as collaborative learning, self-directed learning, experiential-based learning, and active learning, amongst others, have emerged.

In understanding the concept and essence of graduate employability, several theories across various disciplines have been discussed extensively. Among these philosophies is the consensus theory. This theory is centred on the belief that encouraging skills' development among the human capital will certainly guarantee graduate employability and fast-tracked career growth at the workplace (Becker, 1964; Selvadurai *et al.*, 2012). Hence, in the realization of this book, the human capital theory is adopted as the guiding theoretical framework. The justification for its adoption is that this study is essentially concerned with equipping built environment students as they are the future of the construction industry.

This study assumes that employability skills enable individuals to become more flexible and adaptable to the varying demands of the industry, which results in increased productivity to themselves and their employers. This contextual assumption is supported by Hitt *et al.* (2001) and Nafukho *et al.* (2004). At the core of this theory is the notion that skilled individuals are an important production factor that is worth investing in to yield dividends to themselves as well as to the immediate society in which they find themselves (Jackson and Chapman, 2012; Aliu and Aigbavboa, 2017). Thus, Van Loo and Rocco (2004) refer to human capital as an 'investment in knowledge and skills'. Kleynhans (2006) supports this perspective by stating that 'human capital can provide a country with a competitive edge that could lead to economic growth and enhance everyone's welfare'.

Several researchers have described human capital as a multidimensional concept. However, this study employs a critical approach to provide a better understanding of the theory. According to Ben-Porath (1967), Garavan *et al.* (2001), and Omar *et al.* (2012), human capital describes the ability of individuals to display adequate knowledge as well as exhibit competencies in carrying out industry tasks efficiently and effectively. Also, Ben-Porath (1967) asserts that the success of the construction industry is dependent on the successful recruitment, development, and retaining of the best human capital, which is essential for sustaining the future. Benhabib and Spiegel (1994) similarly opine that employee value is highlighted by the skill, knowledge, and experience they exhibit. Hence, employees are considered human capital because they cannot be separated from these values. Moreover, employers are faced with the daunting task of reacting to a continuously changing and fast-paced work setting as the survival of the construction industry remains paramount (Stice, 2011). This scenario has increased the need for more industry-ready graduates, hence placing increased pressure on higher education training systems to deliver (Omar *et al.*, 2012). Another major assumption of the human capital theory is that it emphasizes education as an important stimulator of any given economy (Bridgstock, 2009), and identifies labour as an appropriate tool to carry out production process (Mohr and Seymore, 2012). There has also been evidence that a higher level of training will result in individuals with higher qualifications to meet construction industry needs (Mihail, 2006; Russel *et al.*, 2007; Dacre Pool and Sewell, 2007; Støren and Aamodt, 2010). In addition, Nafukho *et al.* (2004) argue that investment in human capital can positively influence the industry as well as society. Therefore, human capital theorists suggest that an educated and well-informed population is ultimately a productive one. Considering the dynamism of the construction industry, it is expected that human capital be trained to be flexible, adaptable, and skilled to be responsive to the changing demands of industry. According to the human capital theory, industry productivity is guaranteed if a graduate possesses the requisite skills which can lead to both individual and economic benefits. One key question emanating from this discussion is how employability skills can be further developed among built environment students in South Africa and beyond. This book seeks to address this question. This study is of the view that construction education in

South Africa and beyond can be improved and structured in such a manner to meet the employability skills demand of students. This will enable them to acquire the right skills to become more employable when seeking jobs in the labour market. This apparent skills' importance has prompted many governments across industrialized nations around the world to contribute to the positive delivery of employability skills in their educational sectors. This research book therefore argues that the role of the educational sector in employability skills development cannot be overemphasized. For this reason, an employability skill improvement framework will be developed.

Human capital has been adequately discussed across various quarters by several researchers and classical authors. The concept can be dated as far back as the 1960s when Schultz (1961) described the concept as the skills, knowledge, and abilities of employed persons found in an organizational setting. The Schultz (1961) definition was somehow limited as it did not consider the concept of 'value' along with the significance of investing in the resources and efforts of human capital. In 1981, Schultz revised this definition and described human capital as the entirety of human abilities (whether inherent or developed) that are crucial to an organization and can be further developed by a certain level of investment to realize organizational benefits (Schultz, 1981). About 12 years later, 'the health of an individual' (general well-being) was considered as a critical element to include when defining human capital (Becker, 1993). Bontis *et al.* (1999) provided a distinctive definition of human capital. They believe that the abilities, skills, and combined intelligence of the workforce in an organization summarize the essence of human capital in the first place. They posit that individuals who are capable of learning, adapting, and adopting innovativeness can ensure the productivity of an organization on a long-term basis (Bontis *et al.*, 1999). This definition takes into consideration the value of creativity, motivation, innovativeness, dynamism, and adaptability as key tenets when discussing the concept of human capital. Thomas *et al.* (2013) provided a more recent description of the concept of human capital, which focuses on the individuals regarding their performance, output, and inherent potentials in an organizational setting. The inclusion of the term 'potential' indicates the possibility of individuals to further develop their skills and capabilities with the passing of time. This echoes the definition put forward by Dess and Picken (1999), who asserted that the human capital concept consists of the capacity of individuals to improve their existing knowledge, skills, abilities, and experience through continuous learning activities to improve the organizational setting. Their definition establishes 'continuous learning' as a critical construct when discussing human capital. Another dimension of human capital is the element of 'individual productivity' which both Frank and Bernanke (2007) and Acemoglu and Autor (2009) reflected in their definitions. However, Davenport (1999) and Ployhart *et al.* (2014) both acknowledged the element of 'job performance' in their respective definitions. Table 2.1 presents a literature list of some previous research works that offered various definitions of the human capital concept.

Table 2.1 Summary of previous research works in human capital

No.	Author(s) and year	Title of research	Definitions	Variables measured
1	Schultz (1961)	Education and economic growth	Defined as the acquisition of abilities, skills, and knowledge from activities of higher education (training and learning)	Training for the future
2	Mincer (1962)	On-the-job training: costs, returns, and some implications	Described as the required educational process and activities that are instituted by higher education to effectively prepare individuals to handle future organizational responsibilities	Quality workforce for the future
3	Denison (1962)	Sources of economic growth in the United States and the alternatives before us	Defined as intentional educational activities to effectively prepare individuals to handle future organizational responsibilities	Improved workforce
4	Becker (1964)	A theoretical and empirical analysis with special reference to education	Described human capital as an individual's effort in acquiring education	Individual's investment in education
5	Bowman (1969)	Economics of Education	Described human capital as an individual's effort in acquiring education. It also considered several factors that can affect education such as health, social services, etc.	Investment in education
6	Blaug (1976)	The empirical status of human capital theory	Described as the intentional efforts of individuals to acquire holistic education to improve the future	Training for the future
7	Cohn (1980)	The economics of education	Described human capital as a function of improved education and training that can result in better output from individuals as well as higher wages	Increased productivity
8	Romer (1986)	Increasing returns and long-run growth	Defined as an approach for knowledge creation through improved education to improve organizational output	Improved profit

9	Psacharopoulous (1985)	Returns to education: a further international update and implications	Described human capital as a function of improved education and training that can result in better output from individuals	Increased productivity
10	Romer (1987)	Growth based on increasing returns due to specialization	Described the importance of knowledge in human capital discussion	The increasing stock of knowledge
11	Romer (1990)	Endogenous technological change	Took into consideration the total stock of human capital that an organization has. As such, an economy will experience an accelerated growth rate if they possess a larger stock of human capital	Faster rate of growth
12	Becker *et al.* (1990)	Human capital, fertility, and economic growth	Opined that there is a considerable connection between family size and the choice of investing in human capital	Faster economic growth
13	Becker (1993)	Nobel lecture: The economic way of looking at behaviour	Described human capital as an individual's level of education and training that can result in better output. It therefore, referred to the impact of investing in learning and education	Employment and earnings
14	Psacharopoulous and Woodhall (1993)	Education for development: an analysis of investment choices	Described as the intentional efforts of individuals to acquire holistic education to improve socio-economic developments	Investment in education for socio-economic gains
15	Bontis (1996)	There's a price on your head: managing intellectual capital strategically	Defined as the various educational efforts carried out by individuals to meet organizational needs in the long run	Training for the future
16	Huselid *et al.* (1997)	Technical and strategic human resource management effectiveness as determinants of firm performance	Defined as the collective abilities, skills, and knowledge of employees in an organization	Increased productivity

(Continued)

Table 2.1 (Continued)

No.	Author(s) and year	Title of research	Definitions	Variables measured
17	Fitz-Enz (2000)	ROI of human capital: measuring the economic value of employee performance	Defined as the various attributes an individual needs to sustain and thrive in a given job. These include good work ethics, reliability, confidence, commitment, honesty, and a positive attitude	Productivity and efficiency
18	Hitt *et al.* (2001)	Direct and moderating effects of human capital on strategy and performance in professional service firms: a resource-based perspective	Possession of valuable attributes which include education, skills, and experience	Increased productivity
19	David and Lopez (2001)	Knowledge, capabilities, and human capital formation in economic growth	Defined as the various traits and functional abilities that an individual can utilize to great effect	Quality performance
20	Becker (2002)	The age of human capital	Described as the inclusion of the 'health element' (well-being) into the human capital discussion	Health of individuals
21	Youndt and Snell (2004)	Human resource configurations, intellectual capital, and organizational performance	Refers to individual employees' knowledge and skills	Productivity and efficiency
22	Kor and Leblebici (2005)	How do interdependencies among human-capital deployment, development, and diversification strategies affect firms' financial performance?	Refers to an organization's human resources who possess professionals with expert knowledge	Productivity and efficiency
23	Somaya *et al.* (2008)	Gone but not lost: the different performance impacts of employee mobility between co-operators versus a competitor	Refers to the accumulative skills, knowledge, and capabilities of a firm's employees	Productivity and efficiency

24	Coff and Kryscynski (2011)	Drilling for micro-foundations of human capital-based competitive advantages	Described as an individual's capacity to display relevant, skills, and abilities in their chosen profession	Productivity and efficiency
25	Crook *et al.* (2011)	Does human capital matter? A meta-analysis of the relationship between human capital and firm performance	Refers to the skills, abilities, and competencies demonstrated by individuals	Quality performance
26	Ployhart and Moliterno (2011)	The emergence of the human capital resource: a multilevel model	Described as the quality output that can be acquired from individuals who possess job-specific skills and competencies	Productivity and efficiency
27	Wright and McMahan (2011)	Exploring human capital: putting the human back into strategic human resource management	Described as the holistic value of individual attributes that are valuable to an organizational setting	Quality performance
28	Thomas *et al.* (2013)	Human capital and global business strategy	Described as the value of individuals, their abilities and potential to carry out organizational activities	Potentials of individuals to develop skills
29	Ployhart *et al.* (2014)	Human capital is dead; long live human capital resources	Highlighted human capital in the context of organizational outcomes and refers to this type as human capital resources	Deals with three defining elements of human capital resources as regards their structure, function, and level

Source: Author's compilation (2019).

Job market signalling (screening) theory

Unlike the human capital theory, the job market signalling theory is based on the principle that hiring is an investment decision reserved for employers (Arrow, 1973; Spence, 1973; Stiglitz, 1975). It assumes that employers have to make recruitment decisions by taking into account signals which can be conveyed by the educational attainment of job seekers. In another context, signalling theory is also referred to as the 'screening theory', but the underpinnings are the same: job-seeking graduates send signals about their abilities and competencies to employers by obtaining certain academic qualifications, while the employers screen these job applications according to the signals which are transmitted by these academic qualifications (Cai, 2013). Therefore, educational achievements or academic degrees can be regarded as another measure of graduate ability or even quality. Therefore, this theory regards education as a tool for job-seeking graduates to signal their abilities and competencies to employers of labour. This means that the job competencies and innate abilities increase productivity among graduates as against education itself. However, this theory does not provide any connection between the employers and the education providers as well as the learning process (Jonck, 2014).

The concept of employability

Owing to the constant complexities and dynamics of the construction industry, it is becoming more accepted that making career choices and decisions is a lifetime practice (Amundson, 2006). The unpredictable nature of the construction industry ensures that individuals can no longer depend on their academic degrees and credentials alone to find a job or ensure employment security if already employed (Bezuidenhout, 2011). This led Kanter (1989) to suggest a few decades ago that the focus should be shifted from 'employment security' to 'employability security' which enables graduates to be marketable, not only to their current employers, but to other employers in different work disciplines and contexts.

Historical development

Before delving deeper into the employability concept, it is necessary to examine the historical evolution of the concept. Examining its evolution will help to identify and understand the trends of the concept, which helps in achieving the aims of this book. This will provide a platform to arrive at the conceptual framework as well as the selection of suitable research tools. Over the years, there have been several references made to the evolution of the concept (McQuaid and Lindsay, 2005) and its origin can be traced to the 1950s (Sanders and De Grip, 2004; Heijde and Van der Heijden, 2006; Leggatt-Cook, 2007). The historical evolution of the employability concept is discussed next.

Dichotomy employability

This is the first simplistic version of the construct and it appeared in the United States and the United Kingdom from the 1990s to the early 1950s. At the time, the concept was expressed as a dichotomy that was largely connected to able-bodied employees. According to De Grip et al. (2004), a distinction was made between those who could not be employed (elderly and handicapped) and those who could. This was done to provide the unemployable with some form of social assistance and to ensure those employable were absorbed into the labour market (Gazier, 2001). This version of employability was met with several criticisms which included the classification of individuals on the premise of being employable or unemployable, with no degrees of differentiation between them and the non-consideration of the labour market context within this system (Bezuidenhout, 2011).

Socio-medical employability

The socio-medical approach was developed during the 1950s and developed in the United States, the United Kingdom, Germany, and several other countries. At the time, the main focus was on the position of the labour market with regard to the handicapped who were mentally, physically, and socially disabled. Specifically, various aspects such as deficiencies in seeing, hearing, and motor capacity as well as the inability to function intellectually were also considered. According to De Grip *et al.* (2004), attention was paid to the disadvantaged because there was a deficiency of skilled workforce in the post-war period, which led employers to widen their recruitment pool. Hence, this version of the construct rated individuals as more or less employable, taking into cognizance the various deficiencies identified, with a decision to improve employability or to compensate individuals, something which characterized this version (Gazier, 2001).

Manpower policy employability

According to De Grip *et al.* (2004), employability in the 1950s and mid-1960s was regarded as the potential of individuals to gain employment. This version was predominantly developed in the United States, and was an extension of the socio-medical version, but was applied over a much broader population of the unemployed individuals who had problems. The focus here was on both physical and social deficiencies as well as on individual differences such as mobility (regarding the possession of a driver's license) and presentation (with reference to visible drug usage) (Gazier, 2001). The emphasis was based again on the disparities between labour market requirements and a person's skills, knowledge, competencies, and attitudes. This version of employability was solely based on macroeconomic purposes and policymakers who refer to individual attitudes towards employment.

Flow employability

This version of employability was developed predominantly in France as well as other European countries in the 1960s. At the time, the focus was on how certain groups could secure employment. Principally a demand-sided approach, this version of employability emphasized the relative ease with which the jobless are employed within the local and national society (Gazier, 2001). During this stage, employability was described as the expectation and probability of a job seeker actually securing a job (McQuaid and Lindsay, 2005). Sanders and De Grip (2004) argue that from the 1970s, the focus shifted from an individual's 'attitude to work' to 'occupational knowledge and skills'. Factors that influenced this included understanding one's basic occupational knowledge and skills as well as one's possibilities and inherent knowledge about one's position in the labour market and adequate knowledge of what the general employment landscape is like. Towards the end of the 1970s, it was widely accepted that individuals required more than just occupational skills and knowledge to be attractive to employers of labour. It was during this era that the advent of 'transferable skills' came to the fore (Bezuidenhout, 2011).

Labour market performance employability

This version of employability was developed internationally towards the end of the 1970s, and it focused on measurable labour market results. These measures included the probability of securing a job, probable wages, and the probable duration of the employment (Gazier, 2001).

Initiative employability

This version of employability was prominent in the 1980s and it focused on the marketability of individual skills which is measured by social and human capital. Social capital refers to the size and effectiveness of the support network that an individual can utilize, while human capital refers to the skills, knowledge, and learning ability of that individual (Gazier, 2001). This version suggests that the most employable individuals are the ones who can create employment by adopting an entrepreneurial model that happens as a result of harnessing one's skills and connections.

Interactive employability

While the preceding version (initiative employability) was more individually based, the interactive employability embraced a much broader perspective by incorporating both employers and even policymakers as stakeholders in the employability discussion. This version was introduced into the employability debates in the 1990s. It was believed that an individual's employability was relatively connected to the employability of other individuals.

Despite the efforts to elucidate the developmental stages of employability and its meanings, its use in both theory and policies have generated continuous doubt (McQuaid and Lindsay, 2005). The evolutionary view on the employability concept was then described at the societal, industry, and individual levels (Thijssen et al., 2008). They explain that the concept gained ground in the 1970s and was used in solving arising problems; in the 1980s to reshuffle industrial companies to attain effective human resource management; and culminating in the 1990s, as the motive for persons improving successful career prospects in more flexible markets. These conflicting historical accounts of the developmental stages of employability have made them difficult to chronicle. Thankfully, Gazier (2001), McQuaid and Lindsay (2005), and Leggatt-Cook (2007) provide a comprehensive overview of the concept's development which has given rise to the various accepted definitions of the concept over the past century.

Definitions of the concept of employability

Despite the exhaustive discussions on employability over the past few years, the concept remains a contested topic (Gazier, 2001; McQuaid and Lindsay, 2005; Knight, 2006; Hinchliffe and Jolly, 2011; Wilton, 2011). The concept has been widely studied, and its meanings are still debated and vary according to several researchers (Rae, 2007; Wickramasinghe and Perera, 2010; Oliver, 2015; Pitan, 2016). Moreover, the complexity of the concept allows it to be used in various contexts (Pitan, 2016). The literature distilled in this book portrays that employability is not new, dating back to the 1950s; however, its impact was more prominent in the late 1990s (McQuaid and Lindsay, 2005; Knight, 2006; Dacre Pool and Sewell, 2007; Cuyper *et al.*, 2008; Wickramasinghe and Perera, 2010; Finch *et al.*, 2013; Sumanasiri *et al.*, 2015; El Mansour and Dean, 2016). Most notably, the concept has been expanded by the incorporation of components such as company policies, situations, and knowledge of the labour market (Wilton, 2011). In making the concept more relevant to the labour market, there is a need to link employability to an expanded framework of ideas (Hinchliffe and Jolly, 2011). According to McQuaid and Lindsay (2005), this expanded framework is based on the interactivity between the individual and the external environment. Wilton (2011) opines that the concept in a holistic form, with dimensions such as a person's attempt, is to be functional in the desired job, considering the dynamic conditions of the job market. This is supported by McQuaid and Lindsay (2005), who maintain that the concept should be concerned with the individual skills, the circumstances of the individual, and the external environment. The concept in recent times has also been used to refer to a person's employable skills and competencies that are required to thrive in the labour market (Pitan, 2016; El Mansour and Dean, 2016). This perspective resonates with this research because present-day graduates require employable skills to enable them to function effectively in the labour market.

In understanding employability, Rothwell and Arnold (2007) propose an approach based on interrelated components such as the academic performance of students; skills, abilities, and confidence; engagement; ambition; and the perception of universities. Similarly, the psycho-social construct of employability comprises three dimensions, according to Fugate and Ashforth (2004). These are adaptability, career identity, and human and social capital. These constructs propose that employable individuals tend to exhibit higher job satisfaction. They further introduced a new perspective on the concept called 'dispositional employability', which is defined as a collection of individual differences that predispose employees to adapt to their working environment and careers. Dacre Pool and Sewell (2007) explained the meaning of the concept through the development of the 'CareerEDGE' model. They defined this concept as a set of skills, competencies, and personal attributes that makes an individual more likely to choose and remain in a job to achieve success and satisfaction. Another comprehensive definition given by Hillage and Pollard (1998) describes the concept as the capability to be flexible and fulfilled in the labour market. This infers that the concept goes beyond securing a job but fulfilling one's potentials through the use of skills and abilities. The definition which guides this research stems from the one favoured by Knight (2006) and Watts (2006), which describes the concept as 'a set of achievements, skills, understanding and personal attributes that makes graduates more likely to gain employment and be successful in their chosen occupations, which benefits them, the workforce, the community, and the economy'. This definition entails a holistic curriculum that stimulates effective learning to benefit students (Harvey, 2003; Knight, 2006). It also implies that employability is a process of learning which tends to develop key skills and a range of experiences to help students to secure gainful employment. It is also worth noting that the concept does not necessarily represent definite employment; rather, it improves the likelihood of gaining employment. While the extant literature on the concept according to several researchers is so diverse, enhancing graduates for the job market remains the key aim. Table 2.2 presents a literature list of some previous research works that are directly related to the employability concept.

Terms used to describe the concept of employability

According to Weligamage (2009), the concept has been globally described by several other terms. Despite the differences in the labels assigned to these competencies, they all refer to the same meaning: 'key skills', 'graduate attributes', 'life skills', 'soft skills', and 'non-technical skills'. To further understand 'employability', Table 2.3 indicates various conceptualizations of the concept across some developed nations across the globe.

Theories of graduate employability

Two employability theories are considered for the sake of this study: the consensus theory and the conflict theory. Both theories, which provide opposing theoretical bases for the changing relationship between universities, employment, and labour market, date back to the 19th century (Brown *et al.*, 2003). In general, the consensus theory suggests that the integration of generic skills into the university curriculum will improve the employability of graduates and will enable them to function effectively in their positions (Selvadurai *et al.*, 2012). On the other hand, the conflict theory posits that the job of skills development is not fully vested in universities. The conflict theory researcher s argue that employers of labour are also responsible for the development of generic skills. They further opine that employers of labour need to establish collaborations with universities to ensure that students develop key skills which can help them to transit easily into the world of work. Despite the opposing perspectives of both theories, both agree that graduates are to be adequately skilled before entering into the working environment.

Table 2.2 Summary of previous research works in employability

No.	Author(s) and year	Title of research	Definitions	Key elements
1	Hillage and Pollard (1998)	Employability: developing a framework for policy analysis	Defined as being capable of getting and keeping fulfilling work	Capability
2	Fallows and Steven (2000)	Building employability skills into the higher education curriculum	States that possession of skills increases the chances of students securing employment within the construction industry	Building employability skills
3	Overtoom (2000)	Employability skills: an update	Refers to them as transferable core skills that represent essential functional and enabling knowledge required by the present-day industry workforce	Ability to obtain, retain, and adapt to career opportunities
4	Glover and Youngman (2002)	Graduateness and employability: student perceptions of the personal outcomes of university education	Described employability as an enhanced capacity to secure employment, to be familiar with theories on development	Ability to obtain, retain, and adapt to career opportunities
5	Lees (2002)	Information for academic staff on employability	Opines that the concept is multifaceted and there is no commonly accepted meaning	Competencies
6	Brown *et al.* (2003)	Employability in a knowledge-driven economy	Described as the relative chances of acquiring and maintaining different kinds of employment	Acquiring employment
7	Harvey (2003)	On employability	States that it is beyond getting a job but developing the attributes to progress within one's current career	Job satisfaction
8	De Grip *et al.* (2004)	The Industry Employability Index: taking account of supply and demand characteristics	Explains that they are relational skills that are important not only in obtaining a job but also in keeping it and eventually getting another	Retaining employment
9	Forrier and Sels (2004)	The concept employability: a complex mosaic	An individual's chance of landing a job in the internal and/or external labour market	Ability to obtain, retain, and adapt to career opportunities

(Continued)

Table 2.2 (Continued)

No.	Author(s) and year	Title of research	Definitions	Key elements
10	Fugate and Ashforth (2004)	Employability: a psycho-social construct, its dimensions, and applications	Explains it as a psycho-social construct that embodies individual characteristics that foster adaptive cognition, behaviour, and affect and enhance the individual–work interface	Facilitates movements within and between organizations and realization of jobs
11	Greatbatch et al. (2004)	Generic employability skills – aspirations, provision, and perception	Defines them as transferable skills that enhance graduates' capacity to adapt, learn, and work independently	The individual's ability to find and keep a stable job
12	Sanders and De Grip (2004)	Training, task flexibility, and the employability of low skilled workers	The capacity and the willingness to be and to remain attractive in the labour market, by anticipating changes in the task and work environment and reaching to these changes in a proactive way	Multiskilling
13	McQuaid and Lindsay (2005)	The concept of employability	States it as a complex phenomenon that reflects individual characteristics, personal circumstances, and external factors, each of which may affect access to jobs	Building employability skills
14	Cox and King (2006)	Skill sets: an approach to embed employability in course design	The possession of transferable skills which focus on individual attributes that can be exhibited from one job to another, for one's career to succeed	Possession of transferable skills
15	Gazier (2006)	Flexicurity and social dialogue, European ways	It relates to both unemployed people seeking work and those in employment seeking better jobs with their current or a different employee	The individual's ability to find and keep a stable job
16	Rees et al. (2006)	Student employability profiles	Enriches a graduate's foundation of learning and application of subject area	Achievements, skills, knowledge, and personal attributes

17	Heijde and Van der Heijden (2006)	A competence-based and multidimensional operationalization and measurement of employability	Regarded as the continuous fulfilling, acquiring, or creating work through the optimal use of competencies	Competencies
18	Yorke and Knight (2006)	Embedding employability into the curriculum	Defines the concept as a set of skills and which renders an individual relevant to themselves, the industry, and society at large	Achievements, skills, knowledge, and personal attributes
19	Mason *et al.* (2006)	Perception differential between employers and undergraduates on the importance of employability skills	Describes it as a required level of skills and attitudes, students need to possess in securing better jobs in the industry	Achievements, skills, knowledge, and personal attributes
20	Greatbatch and Lewis (2007)	Generic employability skills	Numerated seven domains of employability drawn from various countries	Building employability skills
21	Dacre Pool and Sewell (2007)	The key to employability: developing a practical model of graduate employability	Identifies satisfaction as a key element. The definition stresses the possession of skills and understanding that make graduates more likely to select and secure a job with which they can be content	Job satisfaction
22	Berntson (2008)	Employability perceptions: nature, determinants, and implications for health and well-being'	This refers to an individual's perception of his or her possibilities of getting a new, equal, or better employment	Perception of opportunities
23	Baker and Henson (2010)	Promoting employability skills development in a research-intensive university	Describes it as the chief goal of an educational set-up	Skills development
24	Hinchliffe and Jolly (2011)	Graduate identity and employability	States that the concept is affected by graduate identity. Further elaborates the concept into four categories: values, intellect, performance, and engagement	Building employability skills
25	Torres-Machi *et al.* (2013)	Employability of graduate students in construction management	Observed it as a combination of non-academic skills alongside academic knowledge to solve industry challenges	Ability to obtain, retain, and adapt to career opportunities

(Continued)

Table 2.2 (Continued)

No.	Author(s) and year	Title of research	Definitions	Key elements
26	Oliver (2015)	Redefining graduate employability and work-integrated learning	Suggests that students or graduates can recognize, acquire, and utilize skills in getting industry jobs	The individual's ability to find and keep a stable job
27	Kluve *et al.* (2019)	Do youth employment programmes improve labour market outcomes? A quantitative review	Describes it as a required level of skills and attitudes students need to possess in securing better jobs in the industry	Achievements, skills, knowledge

Table 2.3 Terms used for employability skills in various countries

No.	Country	Terminology used
1	Association of Southeast Asian Nations	Employability skills
2	Australia	Generic skills, key competencies
3	Canada	Employability skills
4	Denmark	Process-independent qualifications
5	Education for All Global Monitoring Report	Transferable skills
6	European Commission	Key competencies
7	France	Transferable skills
8	Germany	Key qualifications
9	International Labour Organization	Core work skills/core skills for employability
10	Latin America	Work competencies, key competencies
11	New Zealand	Essential skills
12	Organisation for Economic Co-operation and Development	Key competencies
13	Singapore	Critical enabling skills training (CREST)
14	Switzerland	Transdisciplinary goals
15	The United Kingdom	Common skills, core skills, key skills
16	The United States	Workplace know-how, necessary skills

Consensus theory

The consensus theory of employability suggests that the development of generic skills at the university level will develop graduates and ensure their readiness as they enter the labour market. According to Fallows and Steven (2000), it is the responsibility of universities to incorporate generic skills into their curricula. These generic skills can be classified into four dimensions: retrieval and handling information (classification and analysis); communication skills and abilities; problem-solving abilities (requisite technical experience and critical thinking skills), and interactive skills and social development (teamwork and responsibility skills). It is therefore the responsibility of universities to integrate these generic skills into their curriculum as academic subject knowledge is not sufficient as a result of the current labour market dynamism. Therefore, Selvadurai *et al.* (2012) state that it is vital for universities to review and adapt their existing curricula periodically to ensure the successful integration of generic skills through various pedagogical methods and approaches. These methods could range from problem-based learning to situated learning.

Conflict theory

The conflict theory of employability suggests that it is the responsibility of employers and universities to jointly develop employability skills (Brown *et al.*, 2003). This theory also identifies the continuous conflict between employers and universities which arises from the dissatisfaction of employers with the universities' effort in providing graduates with the requisite generic attributes and skills (Selvadurai *et al.*, 2012). As in the case of the consensus theory, researchers of this school of thought classify the generic employability skills into three focus areas. The first focus is the higher education context which posits that graduates are to be well skilled in communication, teamwork, and critical evaluating skills. The second area focuses on work placement and experience, which posits that graduates should be given opportunities in the workplace based on their ability to communicate effectively, display knowledge of the subject matter, and exhibit technical experience as well as interpersonal skills. The third context focuses on the graduates' ability to secure employment and maintain it. Because the conflict theory posits that both the employers and universities need to function simultaneously to develop employability skills, it is recommended that their work experience be further honed during their studies. Therefore, universities need to advocate and integrate work (internship) programmes into their curricula, while employers need to ensure ample work placement and opportunities for graduates to develop adequate employability skills and attributes (Selvadurai *et al.*, 2012).

From both theories discussed, it is clear that the consensus theory opines that universities are mainly responsible for the development of generic skills among graduates while the conflict theory places the responsibility on both the employers and the universities. Considering the recent trends of global and economic development, it is essential for both the universities and employers to take responsibility for enhancing graduates' employability skills. However, for the sake of this study, which aims to develop a framework for improving employability, the consensus theory will be thoroughly considered.

Employability models

Across existing literature, there have been numerous employability frameworks and models which have been proposed. As noted earlier, this study focuses specifically on the various ways to improve the employability skills of graduates of the built environment. Based on the various arguments and definitions regarding the concept, it is essential to develop a framework that will increase the understanding of the employability concept, especially in the South African context.

Employability framework by McQuaid and Lindsay

This framework put forward by McQuaid and Lindsay (2005) considers not only individual factors but also external factors and personal circumstances. The underpinnings of these factors are based on the interaction between each of the elements. These viewpoints are not mutually exclusive but ensure that both the 'broad' and the 'narrow' can be combined. From a broad perspective, they argue that the context refers to the market conditions, as well as other external factors that impact an individual's employability (McQuaid and Lindsay, 2005). They also opine that the different skills that an employer may require depend on the dynamism of the environment in which they find themselves. On the other hand, the narrow perspective places emphasis on the individual's attributes and skills set, whilst overlooking other fundamental aspects such as family-friendly policies, easy access to suitable transport, lack of suitable childcare provision, and other aspects which may impede the efforts of a skilled individual from obtaining the job, thereby constraining their employability (McQuaid and Lindsay, 2005). Hence, this model contends that the concept of employability is an interaction between an individual and other external factors.

A heuristic model of employability

Fugate and Ashforth (2004) proposed a framework that is contingent on an individual obtaining the required skills, knowledge, characteristics, and adaptability that employers of labour require. They defined employability as 'a psycho-social construct that embodies individual characteristics that fosters adaptive cognition, behaviour, and effect, and enhance the individual-work interface' (Fugate and Ashforth, 2004). On the basis of this definition, they developed a heuristic model. This model is centred on the concept of individual employability which involves developing several person-centred constructs that are significant for functioning effectively in the world of work. The model, which originates from individual abilities and employer perspectives, can also be referred to as a human resources model. They suggest that the three dimensions of the model (personal adaptability, career identity, and human capital) are imperative. These dimensions represent the willingness, aptitude, and competence to adapt to change, which are key underpinnings of employability. They further maintain that the concept of employability is a form of a work-related trait that facilitates individuals to make their career choices and decisions. Personal traits and skill sets such as self-confidence, dispositions, self-efficacy, long-life learning, and openness, coupled with the affective and cognitive stages of the individual, strengthen their abilities to find and maintain a job (Fugate and Ashforth, 2004). They further maintain that the possession of these traits reduces anxiety and uncertainty and promotes adaptation, which results in performance and satisfaction on the job.

Personal adaptability also implies the ease of responding to various environmental changes and dynamics (Savickas, 1997). Within the adaptability dimension of the Fugate model, several personal constructs are integrated, namely openness, tendency to learn, optimism, and self-control, which help individuals to identify and realize work opportunities. Career identity represents constructs such as role, occupational and organizational identity which refers to how individuals express themselves in a definite job context. According to Arthur *et al.* (1995) and McArdle *et al.* (2007), career identity can be regarded as 'knowing-why' competencies which include personal values, career motivation, and personal meaning. The human and social capital dimensions of this model deal with the personal constructs which can affect the advancement of an individual's career such as educational background, training, age, skills and knowledge, and even work experience (Fugate and Ashforth, 2004; McArdle *et al.*, 2007). Both the human and social capital dimensions are encompassed in the notion of 'knowing-how' career meta-competency (Arthur *et al.*, 1995). Hence, human capital refers to capabilities, explicit and tacit knowledge (Arthur *et al.*, 1995). On the other hand, social capital refers to several relationships formed by an individual as a result of his or her organization's networking activities. According to Symington (2012), this model is not specific to graduates or even students; however, it is vital for understanding the interaction between the stated ideas and components. In further understanding the interaction between the constructs mentioned previously, a dispositional model of employability was proposed by (Fugate and Kinicki, 2008). This model is to nurture adaptive behaviour and individual characteristics to attain positive employment outcomes. The components of the model include career flexibility, proactivity, changes at work, motivation, and work identity. Symington (2012) contends that the model was not developed for graduates, but rather emphasizes worker adaptation. It enhances workers' transition in their workplace and addresses issues of redundancy, redeployment, and even retrenchment.

Van Dam's employability orientation process model

This model of employability takes a different approach from the ones already discussed so far. In this version, Van Dam (2004) considered the antecedents and consequences and classified them as 'orientation' using a process model. According to Van Dam, employability orientation is described as 'the attitudes of employees toward interventions aimed at increasing the organization's flexibility through developing and maintaining workers' employability for the organization' (Van Dam, 2004). In this model, both individual traits (characteristics) and perceptions of employment circumstances are key antecedents of employees' attitudes with regard to development actions and career changes. In this model, two personality characteristics were considered as significant pointers to employability orientation. They are openness (being willing to adopt novel information and significant changes) and initiative (being independently sound and proactive) (Van Dam, 2004; Bezuidenhout, 2011). Another employability orientation is 'tenure' which reveals that low-tenure job seekers seem to get more mobility opportunities and tend to accept these opportunities more voluntarily, have higher expectations regarding mobility, and even have more willingness to participate in employability interventions (Van Dam, 2004; Bezuidenhout, 2011). Two precursors that are work related are career development and perceived organizational support. Career development support is the extent to which supervisors support and develop their subordinates to improve their careers to be beneficial to themselves. On the other hand, perceived organizational support refers to employees' general beliefs about how the organization or company values their contributions and how they show concern about their well-being (Van Dam, 2004; Bezuidenhout, 2011). Van Dam (2004) further stated that employees' career anchors and organizational commitment are key connectors between employability orientation and personal variables. From the model, the career anchors of managerial competence, variety, technical competence, and security are expected to form a stable relationship between the personality variables of initiative and openness and employability orientation. Similarly, from the model, continuance commitment is expected to mediate the relationship between the tenure and the employability orientation. Likewise, affective commitment is expected to mediate the relationship between perceived organizational support and employability orientation (Van Dam, 2004). The model also indicates that there are consequences of employability orientations which give way to employability development activities, such as job experiences and increased knowledge base (Van Dam, 2004). Although it is important to know how employability develops and what its outcomes are, this model is quite different from the previous ones examined by this study. The work-related dimension depicted in this model is not overly relevant to this present research. Van Dam's (2004) model does, however, propose the initiative and openness construct of employability which some other models consider important in conceptualizing and underpinning employability, for example, that of Fugate and Ashforth (2004).

Bridgstock's conceptual model of graduate attributes for employability

Bridgstock's (2009) model illustrates a conceptual framework of graduate attributes that contribute to employability. The dimensions of the model are self-management skills, career-building skills, career management, generic skills, employability skills, discipline-specific skills, and underpinning traits and dispositions. Self-management skills refer to the individuals' perception of themselves in terms of abilities, interest values, and goals. These proficiencies are closely linked to the career identity concept (Arthur *et al.*, 1995; Fugate and Ashforth, 2004) which is a perceived connection between the individuals and their career success. Eby *et al.* (2003) confirmed that students who possess a well-developed concept of their career goals and a true understanding of their abilities possess higher levels of employability than students who do not. Career-building skills include those skills and abilities that are necessary for discovering and implementing information about careers, the labour markets, and the employment world and then locating, securing, and maintaining a job. Career-building skills further provide individuals with industry knowledge to be able to identify opportunities and threats which exist as well as the critical factors which can lead to success (Bridgstock, 2009). They further involve knowing the 'rules of the game' such as the industry norms, values, culture, beliefs, salary scales, employment rates, and details amongst others; possessing the knowledge of identifying the best opportunities to advance with reference to roles within the projects as well as location of the projects; possessing the knowledge of how long to stay in a specific role and when to move on once another opportunity surfaces; and possessing the knowledge of job application. They also involve the ability to display one's skills to be attractive enough to prospective employers or clients as well as possessing the knowledge of how to establish strategic partnerships (both professional and even personal) with people who are well placed to offer possible opportunities (Bridgstock, 2009).

According to Bezuidenhout (2011), career management can be described as successfully attaining generic and discipline-specific skills in the world of work through a continuous process of drawing on reflective, evaluative, and decision-making processes. Career management activities include creating insightful career goals, taking part in valuable learning opportunities, and realizing the connection between the employment landscape, the economy, and the broader society (Bridgstock, 2009; Bezuidenhout, 2011). Generic skills denote key competencies, transferable skills, and employability skills as discussed throughout the study. Bridgstock (2009) states that just a handful of studies have successfully demonstrated that well-developed generic skills can lead to enhanced graduate employability. Employability skills are those skills that directly help individuals to obtain and maintain employment (McQuaid and Lindsay, 2005). In this model, employability skills are divided into two: career management skills and generic/discipline-specific skills. However, career management skills are further divided into two: self-management and career-building (Bridgstock, 2009). On the other hand, discipline-specific skills refer to those skills which are conventionally embedded in higher education curricula to address particular job-related requirements (Bridgstock, 2009). Underpinning traits and dispositions are described as the antecedents that underline the successful applications of career management skills to a successful extent (Bridgstock, 2009). These traits and outlooks include initiative, sociability, career self-efficacy, intrinsic motivation, and self-confidence.

Coetzee's psychological career resources model

In this model, individuals within the contemporary career context are regarded as 'competency traders'. Their attributes, which include knowledge, transferable skills, experiences, and achievements, are all vital for their employability (Coetzee, 2008). Coetzee and Roythorne-Jacobs (2007) view employability as the ability to gain access to, adjust to, and be productive in an employment situation. To successfully adapt to the working environment to attain career success, individuals need to channel their psychological career resources (meta-competencies). Career meta-competencies are defined as 'skills and abilities such as behavioural adaptability, identity awareness, sense of purpose, self-esteem and emotional intelligence, which enable people to be self-directed learners and pro-active agents in the management of their careers' (Coetzee, 2008). Individuals who possess meta-competencies attain specific skills and competencies which improve their general employability as well as occupational expertise. Hence, psychological career resources are viewed as the possession of attitudes, values, abilities, and career-related orientations that improve the general employability of individuals (Coetzee, 2008).

This model is further categorized into five major dimensions: career drivers, career enablers, career harmonizers, career preferences, and career values. For career drivers, the term 'career purpose' is predominant, and it refers to the individual's sense of pursuing a career. Wrzesniewski *et al.* (1997) suggest that individuals who feel strongly called and connected to their careers record higher levels of job satisfaction than those with mere job orientation. Similarly, individuals with a strong sense of purpose tend to achieve job fulfilment and believe that their inputs positively impact society (Bellah *et al.*, 1996). Career drivers include career commitment, career maturity, career motivation, career self-efficacy, and career self-management skills. Career enablers, on the other hand, are the various transferrable skills, self-knowledge, self-concept, work engagement, and career identity which help individuals to be successful in their chosen careers. Career harmonizers refer to the various psychological attributes that help an individual to display flexibility and resiliency. Those attributes include behavioural adaptability, self-esteem, and emotional literacy. They also help to balance the career drivers to prevent individuals from wearing themselves out during career reinvention. Generally, career preferences and career values help individuals to channel their career decisions. From the foregoing discussion, this model sufficiently encompasses several important attributes that are necessary for graduate employability. The model includes constructs such as career identity, adaptability, and social and human capital, which resonates with the model by Fugate and Ashforth (2004). The model also includes constructs such as self-management skills, career self-management skills, and career-building skills which were corroborated by Bridgstock's (2009) model.

USEM model

This can be described as 'a ground-breaking model that can easily be implemented in institutions to make low-cost, high-gain improvements to students' employability' (Yorke, 2006). USEM is an acronym of the components of the model: U represents Understanding subject and knowledge; S represents Skills; E represents Efficacy beliefs, and M represents Metacognition. This model was formed to present a theoretical viewpoint on the concept of employability by highlighting the four broad components of employability skills and the relevance of embedding employability into the HEI curriculum. The four elements highlight what a graduate should acquire from a degree course to be functional in the world of work. These elements are interrelated, according to Yorke (2006). The metacognition aspects and the efficacy belief parts of the model combine to provide the integrated development of applied skills through the curriculum. The model encourages individuals to develop a range of successes and allows universities to adjust their curriculum, unlike other models which speak less on it. The 'U' represents the understanding of the academic subjects; 'S' represents the practice of generic and subject-specific skills; 'E' represents efficacy beliefs that reflect the willingness to act with regard to how individuals respond to new situations. It is an element that refers to various aspects of personality (Knight, 2006); 'M' represents metacognition which refers to the awareness of what one is aware of and the ability to learn more to further improve himself or herself. The four components support the development of employability, making the concept appealing to academics.

DOTS model

This model evolved from career development learning, according to Law and Watts (1977), Watts (2006) and Dacre Pool and Sewell (2007). The model consists of Decision learning (decision-making skills); Opportunity awareness (displaying knowledge of possible work opportunities and requirements); Transition learning (which includes job-search and self-presentation skills), and Self-awareness (which refers to interests and consciousness of one self). This model represents career development learning outcomes. The model emphasizes the employability skills which are necessary for students to possess in order to sell themselves (Watts, 2006). This model is relevant to this study because it involves the career development–learning component which students combine with their ability in identifying opportunities as well as marketing themselves. This model requires students who can make and take decisions in the context of life planning, relate self-awareness to the knowledge of different opportunities, and evaluate how personal priorities may impact upon future career options. It also includes how students devise a career development action plan and identify ways to address opportunities in career development (Law and Watts, 1977). It is pertinent that students are aware of the intricacies of the world of work in order to be able to choose a suitable career path. Opportunity awareness requires students to possess adequate knowledge of various trends in employment and opportunities in one's discipline as well as understanding the requirements of employers (Law and Watts, 1977). The eventual goal of students who undergo construction education is to ease into the job market successfully after graduation. The transition learning element requires students to display a robust understanding of relevant opportunity-search strategies; exhibit the ability to utilize information with regard to job vacancies; detect the obstacles that hinder success in obtaining relevant strategies and opportunities to mitigate them; demonstrate the capacity to adjust oneself to the requirements of opportunities; and finally, possess the necessary skills to effectively navigate job interviews successfully (Law and Watts, 1977). The self-awareness element is pivotal in career development learning and includes the ability to identify abilities, knowledge and transferable skills improved by one's academic achievement; personal skills and how they can be utilized; values and interests with regard to life planning and vocational ideas; strengths and weaknesses and areas which require the need for improvement as well as the channelling of one's strengths and goals to further improve one's profile (Law and Watts, 1977). This component also reinforces students' ambitions in choosing their discipline as well as understanding the reason for their choices. The element encourages self-assessment, personal development planning (PDP), and reflective writing.

Bezuidenhout's graduate employability model

This model by Bezuidenhout is instrumental in employability discussions, not only in the South African context but also to graduates in both developing and developed countries. It emphasizes the various skills and dispositions graduates require to function effectively as well as adapt in the world of work. Building on the work by Fugate and Ashforth (2004) and Fugate and Kinicki (2008), Bezuidenhout (2011) describes employability as 'a psycho-social construct representing a combination of attributes (dispositions, values, attitudes and skills) that promote proactive adaptability in changing environments and enhance an individual's suitability for employment and the likelihood of obtaining career success' (Bezuidenhout, 2011). In this model, several significant dimensions which influence graduate employability are identified. They include career self-management, cultural competence, and underlying dispositions for employability. These dimensions underpin the development of discipline-specific skills, generic skills, and human capital.

The career self-management dimension refers to the ability and tendency of managing one's career. This can be achieved proactively by constantly accumulating career-related information to improve self-knowledge and competencies. According to Reitman and Schneer (2008) and Symington (2012), career self-management is the conception and implementation of one's opportunities to transit easily into the world of work. An important aspect of this dimension is the identification and understanding of one's values and career goals (Quigley and Tymon, 2006). Self-management also relates to career identity which includes individuals' perception of themselves with regard to their interests, abilities, and values (Bridgstock, 2009). Two other concepts relating to career self-management are career-building skills and career development learning. The former are the abilities and knowledge used to seek and implement career-related information in obtaining employment. They also involve the successful utilization of career opportunities to attain favourable career results. These can be achieved by identifying the opportunities and threats within an individual's career choice, identifying the best opportunities to succeed, identifying the best opportunities to exploit a new opportunity, and creating social capital (Bridgstock, 2009). Career development learning, on the other hand, refers to activities that stimulate career knowledge among graduates, which helps them become more attractive to prospective employers (Dacre Pool and Sewell, 2007). The cultural competence dimension refers to an individual's ability, knowledge, and skill to connect with and understand other people of different cultural orientations or ethnic groups (Johnson *et al.*, 2006). This dimension includes knowledge about rules of interaction, values, language, cultural differences, and customs. This dimension comprises two aspects: skills dimension and personal attributes dimension. The skills dimension includes adaptability to different cultural norms and conflict resolution, amongst others. The personal attributes dimension includes traits such as flexibility, willingness to learn, and self-efficacy, amongst others. Greenhaus *et al.* (2010) state that career success relies on one's capacity to understand a multicultural environment successfully. This implies that cultural intelligence is significant in today's workplace.

There are several personal dispositions which are interwoven to promote adaptability in the world of work. The successful interaction between these dispositions can lead to career success and hence improve employability. Dispositions are regarded as stable tendencies for individuals to exhibit certain patterns of behaviours in a variety of situations. They can be referred to by means of various terms such as abilities, habits, and traits (Reber and Reber, 2001). This dimension relates to sociability, entrepreneurial orientation, proactivity, career resilience, openness to change, emotional literacy, and career-related self-evaluations (Bezuidenhout, 2011). Sociability determines whether individuals are open to establishing and maintaining social relationships. It is the ability to use both formal and informal networks to build one's chosen careers. Networks are social behaviours in which an individual seeks different business contacts, and when successful, can influence career outcomes such as income, increased job opportunities, and promotions. Successful networking can also lead to business leads, career advice, and gaining valuable information (Forret and Sullivan, 2002). Entrepreneurial orientation determines whether individuals can utilize their skills and connections to create employment which can improve their employability (Sanders and De Grip, 2004; Moreland, 2006). Gow and McDonald (2000) posit that individuals with entrepreneurial orientation possess the knowledge on how to network effectively and market their ideas as well as themselves. Proactivity refers to various anticipatory activities in which individuals participate to impact themselves or the work environment. Kim *et al.* (2009) contend that proactive individuals are efficient as they look for innovative ways to get results, achieve goals, and improve performance. They can also prepare for the future demands of the labour market and are rarely affected by sudden changes and dynamics in the workplace (Mihail, 2008). According to Van Veldhoven and Dorenbosch (2008), they can actively engage in problem-solving and search for new and inventive ways to achieve greater output (developmental proactivity). Furthermore, Crant (2000) relates proactivity to organizational innovation, entrepreneurship, innovation, leadership, career management, and career success. This means that it is extremely significant for individuals to be proactive when honing their careers to be employable and adaptable to the ever-changing world of work. Career resilience refers to the capacity to recover almost immediately from difficult situations and the ability to be adaptable, flexible, and competent despite encountering adversity (Walker *et al.*, 2006). Individuals who possess resilience have positive self-evaluations and tend to exhibit confidence in handling future challenges (Fugate and Kinicki, 2008). Likewise, Schreuder and Coetzee (2006) describe career resilience as the capacity to adapt to ever-changing circumstances by displaying a readiness to accept organizational changes and being willing to take risks. Hence, career resilience is key for career success as it plays a significant part in career-self management. Openness to change refers to the willingness of individuals to search for new experiences, opportunities, and ideas (Van Dam, 2004). Openness to change therefore enables individuals to consider and gather new information that could help realize career success. Such individuals exhibit adaptability tendencies, which ultimately lead to improved employability (Fugate and Kinicki, 2008).

Emotional literacy refers to the extent to which individuals can recognize, understand, and manage their own emotions and those of other people. It is based on the concept of emotional intelligence (EI), which was first established by Salovey and Mayer (1990). According to the definition, EI is defined as an individual's ability to understand and manage his or her own emotions as well as those of others and to display stable behaviours in solving problems. It also refers to the ability to reason thoroughly about emotions and how it improves the thought processes of an individual (Mayer *et al.*, 2008). Various studies have also pointed out that individuals with high levels of EI experience better career success, build stronger relationships (work and personal), and become more effective leaders (Cooper, 1997; Dulewicz and Higgs, 1998; Cartwright and Pappas, 2008). EI is also related to behavioural adaptability, career progression, career decision-making self-efficacy, and career decision-making and self-confidence (Goleman, 1998; Brown *et al.*, 2003; Coetzee, 2008). Career-related self-evaluation (CSE) refers to a higher order skill which encompasses locus of control, self-esteem, emotional literacy, and generalized self-efficacy. CSE also relates to the basic evaluations that individuals carry out in terms of realizing their career success. Bezuidenhout (2011) suggests that individuals who are higher in CSE tend to evaluate situations more positively, exhibit higher motivation levels, and display a greater level of confidence than individuals who are not. Furthermore, CSE relates to several measures, including increased job and life satisfaction, better work motivation, and even higher income (Judge and Hurst, 2007).

CareerEDGE model

Another relevant model peculiar to this study is the Dacre Pool and Sewell's CareerEDGE model because it improves on the USEM and DOTS models. This model was developed to practically explain the concept of employability, which has been described as indefinable, particularly to students and their parents (Dacre Pool and Sewell, 2007). The acronym 'CareerEDGE' was developed for easy recall; it indicates that, to be employable, all five elements need to be developed by the student. This will result in the development of self-confidence, self-efficacy, and self-esteem, which are key criteria for employability. This model is premised on the definition that 'employability is having the set of skills, knowledge, understanding and personal attributes that makes a person more likely to choose and secure occupations in which they can be satisfied and successful' (Dacre Pool and Sewell, 2007). Dacre Pool and Sewell contend that this model combines the earlier works by Hillage and Pollard (1998), Harvey (2003), and Yorke (2006) into a more logical model which highlights the essential issues that support the concept of employability. The integration of several components of the USEM into the CareerEDGE model reflects the importance of the model. To further clarify this model for students, parents, and non-experts, a 'key' symbol was illustrated by Dacre Pool and Sewell (2007). The key depicts that a student who is equipped with all the elements of the CareerEDGE model can become employable.

CAREER DEVELOPMENT LEARNING

For students to stand a good chance of securing their dream jobs and functioning effectively in those jobs, it is pertinent they receive some education in career development learning to provide career advice, guidance, and directions. As depicted on the DOTS model, career development learning should include various activities which will help students to make the right career choices, explore the industry fronts, identify possible opportunities, and make themselves more attractive to employers (Dacre Pool and Sewell, 2007). The provision of this information to students is one of the critical duties of the career service units of higher education which invariably improves the employability of graduates. It is therefore significant that higher education exposes students to the essence of career education as early as possible (Dacre Pool and Sewell, 2007).

EXPERIENCE: WORK AND LIFE

Work experience is a pivotal element of this model. Generally, students who undergo work-related activities during their academic programmes are presented with the first-hand experience and understand the relationship between theory and practical. According to Jackson (2016), the connection between work experience and employability is established through the inclusion of work-related activities. Jackson suggests that employers place premium value on graduates with hands-on experience when recruiting. The responses from students and graduates who undergo work-related activities further confirm that such experience provides an all-inclusive understanding of the knowledge acquired during classroom settings (Lowden *et al.*, 2011). Work-related activities further provide graduates with a more rounded understanding of the various expectations and intricacies of their chosen careers and consequently lead to improved confidence in their chosen careers. It is therefore widely accepted that graduates who undergo work-related experience are more likely to be more employable than graduates who do not have such opportunities.

DEGREE SUBJECT KNOWLEDGE, UNDERSTANDING, AND SKILLS

These components are central to the model as they encompass the various knowledge and discipline-specific skills which students have to possess upon graduation. It is vital to note that employers generally rate the quality of graduates based on how they successfully complete their academic courses or disciplines because it is primarily the measure available for making recruiting decisions. However, despite possessing the subject-specific knowledge and acquiring an academic degree, these alone cannot successfully determine the employability of a graduate in the eyes of the employers of labour (Dacre Pool and Sewell, 2007).

GENERIC SKILLS

According to several literature sources, generic skills are referred to as 'key skills', 'essential skills', 'core skills', 'soft skills', and 'transferrable skills'. For the sake of this model, the term 'generic skills' will be adopted. With various complexities and dynamics influencing the construction industry, employers are constantly seeking graduates who not only possess an academic degree but are equally equipped with generic skills and technical competencies in several areas. Some of the skills industry employers identify as important include communication, decision-making, analytical, interpersonal, teamwork, information technology, adaptability/flexibility, organizing, creativity, adaptability, and numeracy. These skills, attributes, and competencies go beyond the academic realm and directly influence the employability of graduates (Dacre Pool and Sewell, 2007).

EMOTIONAL INTELLIGENCE (EI)

Various studies have supported the inclusion of emotional intelligence into employability discussions. It has been described as the capability to recognize one's feelings as well as the feelings and thoughts of others (Goleman, 1998; Coetzee and Beukes, 2010). It also refers to the capacity to manage one's emotions and relationship abilities. Studies have shown that individuals who thrive during interviews and develop good working cordiality and relationships are those who can understand their emotions and those of others. Other researchers have shown that students who possess a high level of EI can motivate themselves as well as others, hence leading to successful careers (Dacre Pool and Sewell, 2007). Individuals who possess a high level of EI were found to experience an increased level of confidence in exhibiting their skills and behaviours, according to a study by Coetzee and Beukes (2010). Similarly, Cooper (1997) opines that individuals with high levels of EI enjoy more success in their chosen careers as well as build stronger personal relationships than those with low levels of EI. Jaeger (2003) contends that EI can be promoted in universities through the teaching and learning environment and is positively connected with academic achievement. As with the previously mentioned components of this model, EI is a vital element of employability discussions.

While students are provided with the opportunities to gain the necessary skills and knowledge, it is also pertinent that they are provided with the opportunities to reflect on and evaluate the learning processes that have already taken place. These opportunities provide students with an idea of how far they have developed their employability as well as ways to further develop it. According to Lees (2002), reflection is described as the thoughts of one's actions and how it has influenced the person. A key tool which helps out in this context is personal development planning (PDP) (Lees, 2002). PDP helps students by means of adequate planning of experiences to further improve their self-awareness and thereby employability skills, creating a realistic career plan, and understanding how acquired skills can be honed and applied effectively in various scenarios (Dacre Pool and Sewell, 2007).

SELF-EFFICACY/SELF-CONFIDENCE/SELF-ESTEEM

These three elements of the model are closely interconnected and are called the three 'Ss' because they offer a crucial link between the skills, experience, and understanding as well as employability (Dacre Pool and Sewell, 2007). Self-efficacy refers to the intentional and conscious examination of an individual's skills, abilities, and qualities in achieving results. It also involves the assessment of one's strengths, weaknesses, and interests to experience success. Ghayur and Churchill (2017) explain that students who engage more frequently in self-assessment tend to be more employable and are more productive in their various employments. Another component is self-confidence which refers to one's belief in the possession of competencies and abilities needed to sustain employment. Self-esteem, on the other hand, is an individual's overall judgement about one's own general competencies and abilities. Studies have suggested that an individual with a high level of self-esteem possesses the right attitude to succeed, hence it is one of the components of employability (Dacre Pool and Sewell, 2007).

Summary

To understand the various dimensions of the employability concept across various disciplines, this chapter examined the history of the concept as well as several theories and models. Some of the variables peculiar to the examined employability models include generic skills (GS), career development learning (CDL), subject knowledge (SK), discipline-specific skills (DSS), work experience (WIL), emotional intelligence (EI), external factors, and personal circumstances. However, most of these variables were obtained from several disciplines in the body of knowledge such as management sciences, psychology, business, humanities, and several others. However, the employability concept needs to be discussed in the built environment context in the 4IR era, which is one of the major rationale for this book. The next chapter extensively addresses the gaps observed from previous employability models in a bid to satisfy the purpose of this book.

References

Acemoglu, D. and Autor, D. 2009. *Lectures in labour economics*. Boston, MA: MIT.

Aliu, O.J. and Aigbavboa, C.O. 2017. *Upscaling construction education to meet the needs of the Nigerian construction industry* (Master's dissertation, University of Johannesburg, Johannesburg).

Amundson, N. 2006. Challenges for career interventions in changing contexts. *International Journal for Educational and Vocational Guidance*, 6, pp. 3–14.

Arrow, K.J. 1973. Higher education as a filter. *Journal of Public Economics*, 2, pp. 193–216.

Arthur, M.B., Claman, P.H. and DeFillippi, R.J. 1995. Intelligent enterprise, intelligent careers. *Academy of Management Perspectives*, 9(4), pp. 7–20.

Baker, G. and Henson, D. 2010. Promoting employability skills development in a research-intensive university. *Education + Training*, 52(1), pp. 62–75.

Becker, G.S. 1964. *Human capital: A theoretical and empirical analysis with special reference to Education*. New York: National Bureau and Columbia University Press.

Becker, G.S. 1993. Nobel lecture: The economic way of looking at behavior. *Journal of Political Economy*, 101(3), pp. 385–409.

Becker, G.S. 2002. The age of human capital. In E.P. Lazear (Ed.), *Education in the twenty-first century* (pp. 3–8). Palo Alto, CA: Hoover Institution Press.

Becker, G.S., Murphy, K.M. and Tamura, R. 1990. Human capital, fertility, and economic growth. *Journal of Political Economy*, 98(5, Part 2), pp. S12–S37.

Bellah, R.N., Sullivan, W.M., Tipton, S.M., Madsen, R. and Swindler, A. 1996. *Habits of the heart: Individualism and commitment in American life*. Berkeley: University of California Press.

Benhabib, J. and Spiegel, M.M. 1994. The role of human capital in economic development evidence from aggregate cross-country data. *Journal of Monetary Economics*, 34(2), pp. 143–173.

Ben-Porath, Y. 1967. The production of human capital and the life cycle of earnings. *Journal of Political Economy*, 75(4, Part 1), pp. 352–365.

Berntson, E. 2008. *Employability perceptions: Nature, determinants, and implications for health and well-being* (Doctoral dissertation, Psykologiska institutionen). ISBN (978-91-7155-636-3), Printed in Sweden by US-AB, Stockholm 2008, Distributor: Department of Psychology, Stockholm University.

Bezuidenhout, M. 2011. *The development and evaluation of a measure of graduate employability in the context of the new world of work* (Doctoral dissertation, University of Pretoria, Pretoria).

Blaug, M. 1976. The empirical status of human capital theory: A slightly jaundiced survey. *Journal of Economic Literature*, 14(3), pp. 827–855.

Bontis, N. 1996. There is a price on your head: Managing intellectual capital strategically. *Business Quarterly*, 60(4), pp. 40–46.

Bontis, N., Dragonetti, N.C., Jacobsen, K. and Roos, G. 1999. The knowledge toolbox: A review of tools available to measure and manage intangible resources. *European Management Journal, 17*(4), pp. 391–402.

Bowman, M.J. 1969. 7: Economics of education. *Review of Educational Research, 39*(5), pp. 641–670.

Bridgstock, R. 2009. The graduate attributes we have overlooked: Enhancing graduate employability through career management skills. *Higher Education Research and Development, 28*(1), pp. 31–44. doi: 10.1080/07294360802444347.

Brown, P., Hesketh, A. and Williams, S. 2003. Employability in a knowledge-driven economy. *Journal of Education and Work, 16*(2), pp. 107–126.

Cai, Y. 2013. Graduate employability: A conceptual framework for understanding employers' perceptions. *Higher Education, 65*(4), pp. 457–469.

Cartwright, S. and Pappas, C. 2008. Emotional intelligence, its measurement and implications for the workplace. *International Journal of Management Reviews, 10*(2), pp. 149–171.

Coetzee, M. 2008. Psychological career resources of working adults: A South African survey. *South African Journal of Industrial Psychology, 34*(2), pp. 10–20.

Coetzee, M. and Beukes, C.J. 2010. Employability, emotional intelligence and career preparation support satisfaction among adolescents in the school-to-work transition phase. *Journal of Psychology in Africa, 20*(3), pp. 439–446.

Coetzee, M. and Roythorne-Jacobs, H. 2007. *Career counselling and guidance in the workplace: A manual for career practitioners.* Cape Town: Juta and Co.

Coff, R.W. and Kryscynski, D. 2011. Drilling for micro-foundations of human capital-based competitive advantages. *Journal of Management, 37,* pp. 1429–1443.

Cohn, E. 1980. *The economics of education* (No. Ed. Revised). Cambridge, MA: Ballinger Publishing Company, South Carolina University.

Cooper, R.K. 1997. Applying emotional intelligence in the workplace. *Training and Development, 51*(12), pp. 31–38.

Cox, S. and King, D. 2006. Skill sets: An approach to embed employability in course design. *Education Training, 48*(4), pp. 262–274. http://dx.doi.org/10.1108/004 00910610671933.

Crant, J.M. 2000. Proactive behavior in organizations. *Journal of Management, 26,* pp. 435–462.

Crook, T.R., Todd, S.Y., Combs, J.G., Woehr, D.J. and Ketchen, D.J., Jr. 2011. Does human capital matter? A meta-analysis of the relationship between human capital and firm performance. *Journal of Applied Psychology, 96,* pp. 443–456.

Cuyper, N.D., Bernhard-Oettel, C., Berntson, E., Witte, H.D. and Alarco, B. 2008. Employability and employees' well-being: Mediation by job insecurity 1. *Applied Psychology, 57*(3), pp. 488–509.

Dacre Pool, L. and Sewell, P. 2007. The key to employability: Developing a practical model of graduate employability. *Education Training, 49*(4), pp. 277–289. https://doi.org/10.1108/00400910710754435.

Davenport, T.O. 1999. *Human capital.* San Francisco, CA: Jossey-Bass.

David, P. and Lopez, J. 2001. *Knowledge, capabilities and human capital formation in economic growth.* Unpublished research report prepared for the New Zealand Treasury. Retrieved from: www.Treasury.govt.nz/publications/research-policy/wp/2001/01-13/twp01-13.pdf.

De Grip, A., Van Loo, J. and Sanders, J. 2004. The industry employability index: Taking account of supply and demand characteristics. *International Labour Review, 143*(3), pp. 211–233.

Denison, E.F. 1962. *Sources of economic growth in the United States and the alternatives before us*. New York: Committee for Economic Development.

Dess, G.D. and Picken, J.C. 1999. *Beyond productivity: How leading companies achieve superior performance by leveraging their human capital*. New York: American Management Association.

Dulewicz, V. and Higgs, M. 1998. Emotional intelligence: Can it be measured reliably and validly using competency data? *Competency*, 6(1), pp. 28–37.

Eby, L., Butts, M. and Lockwood, A. 2003. Predictors of success in the era of the boundaryless career. *Journal of Organizational Behavior*, 24(6), pp. 689–708.

El Mansour, B. and Dean, J.C. 2016. Employability skills as perceived by employers and university faculty in the fields of human resource development (HRD) for entry level graduate jobs. *Journal of Human Resource and Sustainability Studies*, 4(1), p. 39.

Fallows, S. and Steven, C. 2000. Building employability skills into the higher education curriculum: A university-wide initiative. *Education Training*, 42(2), pp. 75–83. http://dx.doi.org/10.1108/00400910010331620.

Finch, D.J., Hamilton, L.K., Baldwin, R. and Zehner, M. 2013. An exploratory study of factors affecting undergraduate employability. *Education Training*, 55(7), pp. 681–704. http://dx.doi.org/10.1108/ET-07-2012-0077.

Fitz-Enz, J. 2000. *ROI of human capital: Measuring the economic value of employee performance*. New York, NY: AMACOM, a Division of American Management Association.

Forret, M.L. and Sullivan, S.E. 2002. A balanced scorecard approach to networking: A guide to successfully navigating career challenges. *Organizational Dynamics*, 31(3), pp. 245–258.

Forrier, A. and Sels, L. 2004. The concept employability: A complex mosaic. *International Journal of Human Resources Development and Management*, 3(2), pp. 102–124.

Frank, R.H. and Bernanke, B.S. 2007. *Principles of microeconomics* (3rd ed.). New York: McGraw-Hill/Irwin.

Fugate, M. and Ashforth, B. 2004. Employability: The construct, its dimensions, and applications. *Proceedings of the Annual Meeting of the Academy of Management*, OB: J1–J6, Seattle.

Fugate, M. and Kinicki, A.J. 2008. A dispositional approach to employability: Development of a measure and test of implications for employee reactions to organizational change. *Journal of Occupational and Organizational Psychology*, 81, pp. 503–527.

Garavan, T.N. and Murphy, C. 2001. The co-operative education process and organisational socialisation: A qualitative study of student perceptions of its effectiveness. *Education + Training*, 43(6), pp. 281–302. Retrieved from: https://doi.org/10.1108/EUM0000000005750.

Gazier, B. 2001. Employability: The complexity of a policy notion. In *Employability: From theory to practice* (pp. 3–23). London: Transaction Publishers.

Gazier, B. 2006, May. Flexicurity and social dialogue, European ways. *Ponencia presentada en DG EMPL Seminar on Flexicurity*, Place? Date?, pp. 1–19.

Ghayur, K. and Churchill, D.D. 2017. Career success: Navigating the new work environment. *Career Resources* (3).

Glover, D., Law, S. and Youngman, A. 2002. Graduateness and employability: Student perceptions of the personal outcomes of university education. *Research in Post-Compulsory Education*, 7(3), pp. 293–306.

Goleman, D. 1998. *Working with emotional intelligence*. New York: Bantam.

Gow, K. and McDonald, P. 2000. Attributes required of graduates for the future workplace. *Journal of Vocational Education and Training*, 52(3), pp. 373–396.

Greatbatch, D. and Lewis, P. 2007. Generic employability skills II. *A Paper Prepared by the Centre for Developing and Evaluating Lifelong Learning at the University of Nottingham in Collaboration with the South West Skills and Learning Intelligence Module at the University of Exeter.* Exeter: Skills and Learning Intelligence Module.

Greatbatch, D., Murphy, R., Wilmut, J., Lewis, P., Macintosh, H. and Tolley, H. 2004. Generic employability skills: Aspiration, provision and perception. *Final Report to the Gloucestershire Learning and Skills Council from the Centre for Developing and Evaluating Lifelong Learning.* School of Education, University of Nottingham, Gloucestershire Learning.

Greenhaus, J.H., Callanan, G.A. and Godshalk, V.M. 2010. *Career management.* Thousand Oaks, CA: Sage Publications.

Harvey, L. 2003. *Employability and diversity.* Consultado em www2.wlv.ac.uk/webteam/confs/socdiv/sdd-harvey-0602.

Heijde, C.M.V.D. and Van Der Heijden, B.I. 2006. A competence-based and multidimensional operationalization and measurement of employability. *Human Resource Management: Published in Cooperation with the School of Business Administration, The University of Michigan and in Alliance with the Society of Human Resources Management, 45*(3), pp. 449–476.

Hillage, J. and Pollard, E. 1998. *Employability: Developing a framework for policy analysis.* Research Brief 85. London: Department for Education and Employment.

Hinchliffe, G.W. and Jolly, A. 2011. Graduate identity and employability. *British Educational Research Journal, 37*(4), pp. 563–584.

Hitt, M.A., Bierman, L., Shimizu, K. and Kochhar, R. 2001. Direct and moderating effects of human capital on strategy and performance in professional service firms: A resource-based perspective. *Academy of Management Journal, 44*(1), pp. 13–28.

Huselid, M.A., Jackson, S.E. and Schuler, R.S. 1997. Technical and strategic human resources management effectiveness as determinants of firm performance. *Academy of Management Journal, 40*(1), pp. 171–188.

Jackson, D. 2016. Re-conceptualising graduate employability: The importance of pre-professional identity. *Higher Education Research and Development, 35*(5), pp. 925–939. http://dx.doi.org/10.1080/07294360.2016.1139551.

Jackson, D. and Chapman, E. 2012. Non-technical competencies in undergraduate business degree programs: Australian and UK perspectives. *Studies in Higher Education, 37*(5), pp. 541–567. doi: 10.1080/03075079.2010.527935.

Jaeger, A.J. 2003. Job competences and the curriculum: An inquiry into emotional intelligence in graduate professional education. *Research in Higher Education, 44*(6), pp. 615–639.

Johnson, J.P., Lenartowicz, T. and Apud, S. 2006. Cross-cultural competence in international business: Toward a definition and a model. *Journal of International Business Studies, 37*, pp. 525–543.

Jonck, P. 2014. A human capital evaluation of graduates from the Faculty of Management Sciences employability skills in South Africa. *Academic Journal of Interdisciplinary Studies, 3*(6), pp. 265–274. https://doi.org/10.5901/ajis.2014.v3n6p265.

Judge, T.A. and Hurst, C. 2007. The benefits and possible costs of positive core self-evaluations: A review and agenda for future research. In D. Nelson and C.L. Cooper (Eds.), *Positive organizational behavior.* London: Sage Publications.

Kanter, R.S. 1989. The new managerial work. *Harvard Business Review, 67*(6), pp. 85–92.

Kim, T., Hon, A.H.Y. and Crant, J.M. 2009. Proactive personality, employee creativity, and newcomer outcomes: A longitudinal study. *Journal of Business and Psychology, 24*, pp. 93–103.

Kleynhans, E.P.J. 2006. The role of human capital in the competitive platform of South African industries. *SA Journal of Human Resource Management*, 4(3), pp. 55–63.

Kluve, J., Puerto, S., Robalino, D., Romero, J.M., Rother, F., Stöterau, J., Weidenkaff, F. and Witte, M. 2019. Do youth employment programs improve labor market outcomes? A quantitative review. *World Development*, 114, pp. 237–253.

Knight, P. 2006. *Embedding employability into the curriculum*. Learning and employability series, no. 1. York: Higher Education Academy. Retrieved from: http://www.heacademy. ac.uk/assets/documents/employability/id460_embedding_employability_into_the_ curriculum_338.pdf

Kor, Y.Y. and Leblebici, H. 2005. How do interdependencies among human-capital deployment, development, and diversification strategies affect firms' financial performance? *Strategic Management Journal*, 26, pp. 967–985.

Law, W. and Watts, A.G. 1977. *Schools, careers and community*. London: Church Information Office.

Lees, D. 2002. *Information for academic staff on employability*. LTSN Generic Centre. Retrieved from: www.palatine.ac.uk/files/emp/1233.pdf.

Leggatt-Cook, C. 2007. *Health, wealth and happiness? Employers, employability and the knowledge economy*. Auckland: Labour Market Dynamics Research Programme, Massey University, ISBN: 978-1-877355-30-1.

Lowden, K., Hall, S., Ellio, D.D. and Lewin, J. 2011. *Employers' perceptions of the employability skills of new graduates*. London: Edge Foundation.

Mason, G., Williams, G. and Cranmer, S. 2006. Employability skills initiatives in higher education: What effects do they have on graduate labour market outcomes? *Education Economics*, 17(1), pp. 1–30.

Mayer, J.D., Roberts, R.D. and Barsade, S.G. 2008. Human abilities: Emotional intelligence. *Annual Review of Psychology*, 59, pp. 507–536.

McArdle, S., Waters, L., Briscoe, J.P. and Hall, D.T. 2007. Employability during unemployment: Adaptability, career identity and human and social capital. *Journal of Vocational Behavior*, 71, pp. 247–264.

McQuaid, R.W. and Lindsay, C. 2005. The concept of employability. *Urban Studies*, 42(2), pp. 197–219.

Mihail, D.M. 2006. Internships at Greek universities: An exploratory study. *Journal of Workplace Learning*, 18(1), pp. 28–41. doi: 10.1108/13665620610641292.

Mihail, D.M. 2008. Proactivity and work experience as predictors of career-enhancing strategies. *Human Resource Development International*, 11(5), pp. 523–537.

Mincer, J. 1962. On-the-job training: Costs, returns, and some implications. *Journal of Political Economy*, 70(5, Part 2), pp. 50–79.

Mohr, P. and Seymore, R. 2012. *Understanding microeconomics*. Pretoria: Van Schaik Publishers.

Moreland, N. 2006. *Entrepreneurship and higher education: An employability perspective*. Learning and Employability Series. York: Higher Education Academy. Retrieved from: www.palatine.ac.uk/files/emp/1243.pdf [Accessed 22 May 2010].

Nafukho, F.M., Hairston, N. and Brooks, K. 2004. Human capital theory: Implications for human resource development. *Human Resource Development International*, 7(4), pp. 545–551.

Oliver, B. 2015. Redefining graduate employability and work-integrated learning: Proposals for effective higher education in disrupted economies. *Journal of Teaching and Learning for Graduate Employability*, 6(1), pp. 56–65.

Omar, N.H., Abdul Manaf, A., Helma Mohd, R., Che Kassim, A. and Abd Aziz, K. 2012. Graduates' employability skills based on current job demand through electronic advertisement. *Asian Social Science*, 8(9), pp. 103–110.

Overtoom, C. 2000. *Employability skills: An update.* Center on Education and Training for Employment, ERIC Digest No. 220, Report. Retrieved from: www.cete.org/acve/docgen.asp?tbl=digestsandID=105.

Pitan, O.S. 2016. Towards enhancing university graduate employability in Nigeria. *Journal of Sociology and Social Anthropology*, 7(1), pp. 1–11.

Ployhart, R.E. and Moliterno, T.P. 2011. Emergence of the human capital resource: A multilevel model. *Academy of Management Review*, 36, pp. 127–150.

Ployhart, R.E., Nyberg, A.J., Reilly, G. and Maltarich, M.A. 2014. Human capital is dead: Long live human capital resources! *Journal of Management*, 40(2), pp. 371–398.

Psacharopoulous, G. 1985. Returns to education: A further international update and implications. *Journal of Human Resources*, pp. 583–604.

Quigley, N.R. and Tymon, W.G., Jr. 2006. Toward an integrated model of intrinsic motivation and career self-management. *Career Development International*, 11(6), pp. 522–543.

Rae, D. 2007. Connecting enterprise and graduate employability: Challenges to the higher education culture and curriculum? *Education + Training*, 49(8/9), pp. 605–619.

Reber, A.S. and Reber, E. 2001. *The Penguin dictionary of psychology* (3rd ed.). London: Penguin books.

Rees, C., Forbes, P. and Kubler, B. 2006. *Student employability profiles.* Heslington, York: The Higher Education Academy.

Reitman, F. and Schneer, J.A. 2008. Enabling the new careers of the 21st century. *Organization Management Journal*, 5(1), pp. 17–28.

Romer, P.M. 1986. Increasing returns and long-run growth. *Journal of Political Economy*, 94(5), pp. 1002–1037.

Romer, P.M. 1987. Growth based on increasing returns due to specialization. *The American Economic Review*, 77(2), pp. 56–62.

Romer, P.M. 1990. Endogenous technological change. *Journal of Political Economy*, 98(5, Part 2), pp. S71–S102.

Rothwell, A. and Arnold, J. 2007. Self-perceived employability: Development and validation of a scale. *Personnel Review*, 36(1), pp. 23–41.

Russell, J.S., Hanna, A., Bank, L.C. and Shapira, A. 2007. Education in construction engineering and management built on tradition: Blueprint for tomorrow. *Journal of Construction Engineering and Management*, 133(9), pp. 661–668. doi: 10.1061/(ASCE)0733-9364(2007)133:9(661).

Salovey, P. and Mayer, J.D. 1990. Emotional intelligence. *Imagination, Cognition and Personality*, 9(3), pp. 185–211.

Sanders, J. and De Grip, A. 2004. Training, task flexibility and the employability of low-skilled workers. *International Journal of Manpower*, 25(1), pp. 73–89.

Savickas, M.L. 1997. Career adaptability: An integrative construct for life span, life-space theory. *The Career Development Quarterly*, 45(3), pp. 247–259.

Schreuder, A.M.G. and Coetzee, M. 2006. *Careers: An organisational perspective* (3rd ed.). Lansdowne: Juta and Co.

Schultz, T.W. 1961. Investment in human capital. *American Economic Review*, 51, pp. 1–17.

Schultz, T.W. 1981. Education and economic growth. In N.B. Henry (Ed.), *Social forces influencing American education.* Chicago, IL: University of Chicago Press.

Selvadurai, S., Choy, E.A. and Maros, M. 2012. Generic skills of prospective graduates from the employers' perspectives. *Asian Social Science*, 8(12), pp. 295–303.

Somaya, D., Williamson, I.O. and Lorinkova, N. 2008. Gone but not lost: The different performance impacts of employee mobility between co-operators versus competitors. *Academy of Management Journal*, 51, pp. 936–953.

Spence, M. 1973. Job market signalling. *Quarterly Journal of Economics*, 87(3), pp. 355–374.

Stice, J.M. 2011. *A leader's introduction of rapid and radical change into an organization: A case study of Jack Welch and general electric* (Doctoral dissertation, The George Washington University, Washington, DC).

Stiglitz, J.E. 1975. The theory of "screening", education and the distribution of income. *American Economic Review*, 65(3), pp. 283–300.

Støren, L.A. and Aamodt, P.O. 2010. The quality of higher education and employability of graduates. *Quality in Higher Education*, 16(3), pp. 297–313.

Sumanasiri, E.G.T., Ab Yajid, M.S. and Khatibi, A. 2015. Review of literature on graduate employability. *Journal of Studies in Education*, 5(3), pp. 75–88.

Symington, N. 2012. *Investigating graduate employability and psychological career resources* (Doctoral dissertation, University of Pretoria, Pretoria).

Thijssen, J.G., Van der Heijden, B.I. and Rocco, T.S. 2008. Toward the employability-link model: Current employment transition to future employment perspectives. *Human Resource Development Review*, 7(2), pp. 165–183.

Thomas, H., Smith, R.R. and Diez, F. 2013. *Human capital and global business strategy*. New York: Cambridge University Press.

Torres-Machi, C., Dahan, A., Yepes, V. and Pellicer, E. 2013. Comparative study of employability between Spanish and French students in civil engineering. *Journal of Professional Issues in Engineering Education and Practice* (American Society of Civil Engineers, ISSN-1052-3928), 139(2), pp. 163–170.

Van Dam, K. 2004. Antecedents and consequences of employability orientation. *European Journal of Work and Organizational Psychology*, 13(1), pp. 29–51.

Van Loo, J.B. and Rocco, T.S. 2004. Continuing professional education and human capital theory. *Online Submission*, Austin, TX, United States of America.

Van Veldhoven, M. and Dorenbosch, L. 2008. Age, proactivity and career development. *Career Development International*, 13(2), pp. 112–131.

Walker, C., Gleaves, A. and Grey, J. 2006. Can students within higher education learn to be resilient and, educationally speaking, does it matter? *Educational Studies*, 32(3), pp. 251–264.

Watts, A.G. 2006. *Career development learning and employability*. York: Higher Education Academy.

Weligamage, S.S. 2009. *Graduates' employability skills: Evidence from literature review*. Dalugama: University of Kelaniya.

Wickramasinghe, V. and Perera, L. 2010. Graduates', university lecturers' and employers' perceptions towards employability skills. *Education + Training*, 52(3), pp. 226–244.

Wilton, N. 2011. Do employability skills really matter in the UK graduate labour market? The case of business and management graduates. *Work, Employment and Society*, 25(1), pp. 85–100.

Wright, P.M. and McMahan, G.C. 2011. Exploring human capital: Putting human back into strategic human resource management. *Human Resource Management Journal*, 21, pp. 93–104.

Wrzesniewski, A., McCauley, C.R., Rozin, P. and Schwartz, B. 1997. Jobs, careers and callings: People's relations to their work. *Journal of Research in Personality*, *31*(1), pp. 21–33.

Yorke, M. and Knight, P.T. 2004. Learning & employability. *Embedding Employability into the Curriculum*, 3(1–28).

Youndt, M.A. and Snell, S.A. 2004. Human resource configurations, intellectual capital, and organizational performance. *Journal of Managerial Issues*, *16*, pp. 337–360.

3   Gaps in employability research

Evaluation of the selected employability models

This section analyses some of the theoretical models which underpin this book in developing an employability improvement framework. The purpose of this critical evaluation is to assess how the model constructs can be adopted into the South African context, particularly for the employability of built environment graduates. The model for this study synthesizes the seminal contributions of Law and Watts' (1977) DOTS model, the heuristic model of employability of Fugate and Ashforth (2004), Van Dam's (2004) employability orientation process model, McQuaid and Lindsay's (2005) employability framework, Yorke's and Knight (2006) USEM MODEL, Dacre Pool and Sewell's (2007) CareerEDGE model, Bridgstock's (2009) conceptual model of graduate attributes for employability, Coetzee's (2012) psychological career resources model, and Bezuidenhout's (2011) graduate employability model.

These models all possess distinct merits and emphasize several key elements and constructs which adequately contribute to an employability improvement framework for built environment graduates in South Africa. The DOTS model, which is valued for its simplicity, highlights the need for career development learning as a key element in the employability discussion. The acronym represents Decision learning (decision-making skills), Opportunity awareness (displaying knowledge of possible work opportunities and requirements), Transition learning (which includes job-hunting abilities and self-presentation), and Self-awareness (which refers to interests and consciousness of oneself), thereby representing career development learning outcomes (Law and Watts, 1977). The heuristic model of employability suggests that employability is centred on several person-centred constructs which are needed for graduates to possess after graduation. These constructs take into cognizance the elements of human capital, personal adaptability, and career identity (Fugate and Ashforth, 2004). Like the heuristic model, the employability orientation process model also considers the ability of employees to gain and maintain employment (Van Dam, 2004). The employability framework by McQuaid and Lindsay (2005) studies the various influences of certain factors on the employability concept. These elements include individual factors, personal circumstances, and external factors (McQuaid and Lindsay, 2005). The USEM model highlights what a graduate should acquire from a degree course to attain job success. These include the USEM acronyms which represent Understanding subject and knowledge, Skills, Efficacy beliefs, and Metacognition (Yorke and Knight, 2006). This model has been widely considered as one of the most effective and recognized models in the employability discourse. The model also acts as a guide for academicians in the integration of employability into university curricula, hence its adoption for this book.

The conceptual model of graduate attributes for employability by Bridgstock highlights the contributions of self-management skills, career-building skills, career management, generic skills, employability skills, discipline-specific skills, and underpinning traits and dispositions to the employability discussion (Bridgstock, 2009). The psychological career resources model indicates several employability constructs, including career drivers, career enablers, career harmonizers, career preferences, and career values. These constructs constitute several elements, including self-management skills, career self-management skills, career-building skills, and transferable skills, amongst others (Coetzee, 2012). The graduate employability model by Bezuidenhout (2011) highlights several significant dimensions which influence graduate employability, including career self-management, cultural competence, and underlying dispositions for employability. Most significantly, the CareerEDGE model seems to encapsulate several dimensions which apply to the wider and contextual study of the concept of employability. These constructs include the possession of degree subject knowledge, understanding, and skills (discipline-specific skills), career development learning, generic skills, work experience, emotional intelligence, and the 3S constructs (self-efficacy/self-confidence/self-esteem). For graduates to be adequately prepared to fit into the world of work, they need to possess these elements, according to Dacre Pool and Sewell (2007).

Gaps in employability research

The previously mentioned models are legitimate and accepted by higher education institutions (HEIs) across the world and provide varied perspectives on the scope of graduate employability which can be adopted in the built environment. The nine models also contain various constructs that apply to the realization of this study such as generic skills, discipline-specific skills, work-integrated learning, and emotional intelligence (Pool and Sewel, 2007). However, some of these frameworks have become dated because the concept of employability is constantly evolving as present-day graduates need to possess more proficiencies to function effectively in this dynamic era of work, coupled with the recent wave of the 4IR. In addition, the fact that the various model constructs correlate with other models implies that any perceived weakness in any particular model will be addressed. Like several employability studies carried out in South Africa (Griesel and Parker, 2009; Bezuidenhout, 2011; Symington, 2012; Archer and Chetty, 2013; Ndzube, 2013; Potgieter, 2013; Geel, 2014; Jonck, 2014; Mtebula, 2015; Kundaeli, 2016; Gani, 2017; Naicker, 2017; Fongwa, 2018; Shivoro *et al.*, 2018), most of the considered models have discussed and adopted the concept in various fields such as the management sciences, psychology, commerce, economics, and business studies, amongst others.

However, there has been little mention of an employability skills model specifically for the built environment domain in the 4IR era, a gap which this book intends to address. This section attempts to address the two gaps which have been identified, namely the roles of university–industry collaboration, and the roles of Industry 4.0 requirements in the improvement of employability skills among built environment graduates. The consideration of these identified gaps is based on the notion that the present-day construction industry is driven by various factors, including technology and globalization. Hence, universities must establish cordial connections with industry counterparts in a bid to create developmental strategies. Also, in embracing the 4IR, built environment graduates will require a plethora of skills as automation continues to advance.

Gap 1: university–industry collaborations

The idea of university–industry collaboration (UIC) is not new. In fact, it has existed for a long time, principally in technology-driven regions. Collaborations came into prominence in 1903 and were propelled by the Sunderland Technical Knowledge in Northern England and were known as the Sandwich Programme (Ramakrishnan and Yasin, 2011). Since then, several researchers have highlighted the significance of UIC, and this book identifies it as a key component to improve employability skills among built environment graduates. Several terminologies can be used in place of UIC. They include university–industry partnership (UIP), university–industry linkages (UIL), university–industry relationship (UIR), and university–industry alliance (UIA). Although these terms can be used interchangeably, this book will make use of university–industry collaboration (UIC), except where it becomes necessary to adopt a different term. According to Siegel *et al.* (2003) and Bekkers and Bodas Freitas (2008), UIC refers to the alliance and dealings between the higher educational system (universities) and the industry with the hope of achieving both technology and knowledge exchange. Winer and Ray (1994) suggest that UIC is a purposeful and well-explained relationship between two individuals or multiple stakeholders in realizing outcomes that may be difficult to achieve as a separate entity. They further proposed several critical elements in achieving effective collaboration. They insist that collaboration should be mutually beneficial (win-win) for all parties involved; strictly defined (parties should be aware of their specific roles, values, and end game); inclusive of a variety of entities; and purpose-driven (Winer and Ray, 1994).

The motivation for university–industry collaborations

According to Bettis and Hitt (1995) and Wright *et al.* (2008), there are several reasons why both the university and industry seek collaborations. For universities, motives such as growth in new knowledge, improvement in teaching approaches, enhancement of reputation, access to funding, grants, and access to empirical data from industry place significant pressures on universities to encourage collaboration. For the industry, motives such as rapid technological change, recruiting decisions, global competition, development of new products and processes, and accessibility to university's equipment, amongst others, similarly place pressures on them to seek collaboration (Bettis and Hitt, 1995; Hagan, 2004; Wright *et al.*, 2008). These pressures have prompted an increased push for developing effective UICs which enhance innovation and economic competitiveness (Perkmann *et al.*, 2013).

UNIVERSITIES' PERSPECTIVE

Necessity   With a constant rise in global competition, coupled with increased innovative technologies, governments seek partnerships to improve innovations, thereby leading to growth and wealth creation (Barnes *et al.*, 2002). The functioning interface between them can also lead to the well-being of the economy. Hence, universities are constantly looking to encourage UIC in response to government strategies (Perkmann *et al.*, 2011; Ankrah and Omar, 2015).

Reciprocity   While universities offer a wide range of research and development opportunities, the industry offers a variety of opportunities for the development of products, commercialization, and possible employment of graduates after graduation (Sherwood *et al.*, 2004). This can motivate universities to establish rapport with industries to explore these possible benefits (Ankrah and Omar, 2015).

Efficiency   While the government constantly makes funding available for UIC initiatives, universities are always motivated to seek an alternative source of revenue to minimize their dependence on the public purse. This can be achieved through the commercialization of research and the licensing of patents (Logar *et al.*, 2001). Collaborations also appeal to universities because industry funding is less bureaucratic compared to public service interventions (Ankrah and Omar, 2015). Personal financial gain can be a further motivator for faculty members to establish liaisons with industry (Siegel *et al.*, 2004; Ankrah and Omar, 2015).

Stability   According to Boddy *et al.* (2000), the collaboration theory recommends inter-organizational relationship as an approach to steadiness when the economic climate becomes unpredictable. Oliver (1990) also suggests that organizations are motivated by the stability theme and tend to enter into collaboration to mitigate environmental uncertainties. The growth in new knowledge can further motivate universities to establish collaborations with industry to sustain their role as a knowledge hub. These collaborations can provide fertile grounds for university scholars to test theories and hone their skills. Santoro and Chakrabarti (2001) also posit that universities seek to collaborate to provide industrial exposure for students, thereby improving their employability. This implies that collaboration can improve the quality of teaching and enhance curriculum development. Harman and Sherwell (2002) further suggest that universities can be motivated to collaborate because it improves the opportunity to publish journals and high-level scholarly works which would strengthen their original mission in knowledge dissemination (Newberg and Dunn, 2002; Ankrah and Omar, 2015).

Legitimacy   Another fundamental reason why universities seek collaborations could be a deep-down aspiration to expand their frontiers on a bigger scale (Mora-Valentin, 2000). Universities experience various forms of pressures from society (public and political) to demonstrate increased social accountability and overall economic relevance in their functions (Cohen *et al.*, 1998). These pressures therefore stimulate universities to consider collaborations in contributing to economic development via knowledge exchange (Siegel *et al.*, 2003; Hagan, 2004). Siegel *et al.* (2004) also suggest that another motivation for universities is increasing their knowledge credibility and status within the industrial scientific community. This can be attained through joint publications of scholarly works as well as presentations at significant conferences. The funds which can be obtained from industry can be further channelled into more groundbreaking research which enhances their academic reputation (Siegel *et al.*, 2004; Ankrah and Omar, 2015).

INDUSTRY PERSPECTIVE

Necessity   Owing to the rapid global changes combined with the increasingly competitive and technological environment, governments have been compelled to advocate for research interactions in the belief that universities can revive shaky economies if they diffuse and disseminate their knowledge through industrial partnerships (Perkmann *et al.*, 2013). This means several national research programmes can be initiated by governments, and industries can be compelled to collaborate with universities to benefit from these constitutional policies (Ankrah and Omar, 2015).

Reciprocity   Another motivation for industry to seek collaboration is accessibility to the best students for internship purposes and even recruiting possibilities (Siegel *et al.*, 2003). UIC can also provide industry employers with ample access to faculty members, researchers, and even educators on a consultation basis (Perkmann *et al.*, 2011; Ankrah and Omar, 2015).

Efficiency   Another germane motivation for industry to seek collaboration is to attain efficiency and functionality. Cohen *et al.* (1998) believe that research between both parties can enhance the industry's growth index, cost savings, research and development productivity, commercial power, financial windfall, innovative outputs, and even patenting activity (Cohen *et al.*, 1998). These various benefits can provide the industry with a competitive advantage, a serious motivation to seek UIC. Moreover, the access to trailblazing technologies and up-to-date research facilities and expertise can further motivate industry to seek collaboration. These possible benefits and accessibility can help to reduce the distance between design and production as well as mitigate the impact of shorter product life cycles, giving the industry some competitive advantage (Oliver, 1990; Ankrah and Omar, 2015).

Stability   The quest to solve specific complex problems can prompt the industry to establish a partnership to perform academic research. Several studies also demonstrate that collaboration between both parties can stimulate technology-driven firms, which can lead to exponential growth (Ankrah and Omar, 2015). Schartinger *et al.* (2006) suggest that collaboration is appreciated even for firms with a strong capacity to carry out research and development as it reduces risk and maximizes their resources (human and capital). Access to other research networks which can involve numerous firms from different sectors is also a further motivation for the industry to seek collaborations (Oliver, 1990; George *et al.*, 2002; Ankrah and Omar, 2015).

Legitimacy   As in the case of universities, another fundamental motivation for industry to seek collaboration is an intrinsic desire to enhance its prestige on a bigger scale. Siegel *et al.* (2003) suggest that industry (firms and companies) can enhance their corporate image and reputation by associating with a prestigious university. Establishing rapport with reputable and research-driven universities can improve a firm's ratings and legitimacy in the eyes of other influential investors and parastatals (Oliver, 1990; Hong and Su, 2013; Ankrah and Omar, 2015).

Asymmetry   Another fundamental motivation for industry to seek collaboration is to commercialize universities-based technologies for their financial purpose and professional recognition (Siegel *et al.*, 2003). In doing so, the industry can seek exclusive rights to generated technologies and can steer the direction of the research to their benefit (Oliver, 1990; Newberg and Dunn, 2002; George *et al.*, 2002; Ankrah and Omar, 2015).

Outcomes of university–industry collaborations

As in the case of any inter-organizational partnership, collaboration between university and industry is also faced with its pros and cons for all stakeholders involved. These benefits are categorized into three dimensions: economic benefits (positive contributions to the overall economy), institutional benefits (mutual benefits enjoyed by both parties involved), and positive external outcomes or social benefits (positive contributions to the society) (Ankrah and Omar, 2015). By contrast, there is a downside to UIC, according to literature. It is germane for both parties, most especially universities, to be aware of the possible shortcomings so that measures can be taken to mitigate failures and ensure successful collaborations (Harman and Sherwell, 2002). These disadvantages are classified into four categories: deviation from original statements and intentions, quality issues, conflicts, and risks. Both the benefits and downsides of UIC are discussed further.

BENEFITS OF UNIVERSITY–INDUSTRY COLLABORATIONS

Economic benefits   These are referred to as the positive contributions to the overall economy at large. For universities, collaborations provide an avenue to generate income from both private and public parastatals as well as the researchers involved. Through collaboration, discoveries and findings could be patented, leading to several business opportunities that invariably contribute to the well-being of the economy, both socially and economically. Similarly, the industry stands to benefit from collaboration in terms of the creation of innovative products and processes which could earn patents in the long run. Moreover, collaborating with universities could be seen as a cost-effective method of conducting external research rather than embarking on in-house studies. Employers could also benefit from the public grants that come because of collaborations, which can lead to wealth creation as well as economic development (Lee, 2000; Ankrah and Omar, 2015).

Institutional benefits   These are referred to as the mutual benefits enjoyed by both parties involved. For universities, collaborations provide many opportunities for students to experience the rigours, reality, and practicality of industry-related issues. Hence, collaborating with industry is an avenue for curriculum development. It further provides an opportunity for faculty staff and university personnel to obtain valuable feedback from their industry counterparts regarding research ideas and proposed academic theories. This can lead to the joint publishing of scholarly articles (papers and journals) to solve arising problems in an economy. Consequently, collaboration can result in the creation of spin-off firms and affiliated institutes that can foster research and development as well as train students for the future. Moreover, through collaboration, universities can access advanced equipment from their industrial counterparts. Similarly, the industry stands to benefit from collaboration in terms of accessing innovative research ideas and knowledge that emanate from multidisciplinary members of the university setting. Through these innovative ideas, employers can access a broader network of research expertise. It further provides the opportunity for employers to contribute to research trends and initiatives that can benefit their firms in the long run. Collaboration also provides the industry with access to talented prospects and students who are well trained and can add value to the industry. In addition, it provides an avenue for industry professionals to function in a consultancy role to solve arising technical problems. Through collaboration, the industry can enhance their ideas through joint publications of scholarly articles (Lee, 2000; Ankrah and Omar, 2015).

Social benefits

These are referred to as positive contributions to society. For universities, collaborations provide many opportunities for university institutions to provide valuable contributions to the immediate community. Present-day institutions also take pride in establishing affiliations and partnerships with several industrial bodies, thereby enhancing their reputation in the process. Similarly, the industry stands to benefit from collaboration in terms of their reputation as research-driven and trendy, enhancing their reputation in the process (Lee, 2000; Ankrah and Omar, 2015).

DOWNSIDES OF UNIVERSITY–INDUSTRY COLLABORATIONS

Deviation from core ethics   It is normal for universities to design and work towards achieving their mission statement and core objectives. However, in a bid to be result oriented in achieving their short-term goals, they establish collaborations with industry, which could result in deviation from their original statements and long-term objectives. Moreover, the implementation and dissemination of knowledge could be hindered because of drafted confidentiality agreements that may arise as a result of collaborations. There is also the concern that universities could become an extension of industrial research activities on a short-term basis, especially when the industry requires emergency solutions to problems. If this move is commercially motivated, there could be a clash of interest for the university with regard to upholding their mission statement and fulfilling the obligations of collaborations. Owing to the bureaucratic processes that sometimes exist in the academic space, the industry could face difficulties in technology commercialization and delayed execution of their obligations. Moreover, collaborations could prompt the industry to create certain managerial and administrative arms to establish effective collaboration with universities. This can incur extra cost to the industry as well as time delay in decision processes because of the number of parties involved (Lee, 2000; Ankrah and Omar, 2015).

Quality issues   For universities, collaborating with industry comes at a price regarding quality issues. Instead of focusing on their core educational activities, there is the possibility of misplaced priorities of staff or faculty members who are involved in industry interaction. In a bid to meet with the demands and requirements of collaborating, the commitment of university members could be called into question. Collaboration with industry can also alter the quality and quantity of research conducted by universities as they are sometimes obligated to run ideas by their partners, resulting in possible variances as opposed to the original idea. Collaborations can also result in a possible clash of interests because university members are highly theoretical, while their industrial counterparts are more hands-on with regard to solving problems. However, striking a mutual understanding can mitigate this possible problem (Lee, 2000; Ankrah and Omar, 2015).

Conflicts   There is no doubt that both universities and industry seek collaboration to meet their mutual needs effectively. However, in the event of a breakdown in communication or failure to achieve the set out objectives, there is the possibility of conflicts between the parties in realizing the outcomes of such setbacks. Owing to different mindsets and motives during collaboration, the possibility of conflicts exists with regard to patenting and intellectual properties (Lee, 2000; Ankrah and Omar, 2015).

Risks   For universities, collaboration comes with its risks and uncertainties. Sometimes, the university board faces the difficult task of publishing research results to attain scholarly recognition and short-term returns or keeping inventions under wraps pending patenting, with the risk of the technological idea or findings becoming obsolete. For industry, there is the possibility of leakage of confidential information because of the high failure rate of collaborations. Collaborations are also regarded as a financial and market risk for industries because of the uncertainty surrounding the success of launched products (Lee, 2000; Ankrah and Omar, 2015).

Theories of university–industry collaborations

As discussed, collaborations between university and industry can be examined from two perspectives: rational and irrational. From these processes, two theoretical views are obtainable – the interdependency and interaction theories. The interdependency theories examine the influence of external factors and the environment on collaborations, while interaction theories deal with the internal development of the relationship itself (Geisler, 1995).

INTERDEPENDENCY THEORIES

Several perspectives are obtainable in the interdependency theories. These are sociological perspectives, behavioural-oriented paradigms, and economics-oriented paradigms (Geisler, 1995). Similarly, Barringer and Harrison (2000) suggest six perspectives: transaction costs economics, resource dependency, strategic choice, stakeholder theory, learning theory, and institutional theory.

TRANSACTION COST ECONOMICS

According to Tadelis and Williamson (2012), the transaction cost economics (TCE) acknowledges transaction as the fundamental unit of analysis for an organizations' economic relationship. These relationships are required for efficiency maximization and cost-reduction rationale (Tadelis and Williamson, 2012). This may provide some motivation for both universities and firms to collaborate to reduce the cost of their technology development ventures (Dekker, 2004). In simple terms, TCE highlights the ways an organization organizes its activities to cut down on both production and transaction costs (Barringer and Harrison, 2000).

RESOURCE DEPENDENCY THEORY

The resource dependency (RD) theory is embedded in an open-system framework which proposes that firms or organizations must interact with their immediate environment to acquire resources to survive (Scott, 1987). The need to obtain these resources creates a certain level of dependency between organization and external units, which can be governmental agencies, competitors, creditors, suppliers, or any other relevant entity in a firm's environment (Barringer and Harrison, 2000). In managing these dependencies successfully, RD theorists contend that organizations must be able to self-sustain to minimize dependence on their counterparts and acquire resources that would prompt other organizations to depend on them (Pfeffer and Salancik, 1978). In achieving these objectives, participation in inter-organizational relationships is a good approach by firms. This is because an alliance with other firms enables accessibility to critical resources and increases their influence with other organizations. In simple terms, RD theory proposes that a firm may seek an alliance with a firm to gain a significant advantage over a competitor or to simply bridge a skill or resource gap (Mitchell and Singh, 1996; Barringer and Harrison, 2000).

Another common notion of the RD theory is that establishment sought collaborations to benefit from resources not inherent to them. For example, the collaboration between small subcontracting firms and larger engineering firms enables a win-win for both parties. In this case, the larger firms benefit from the collaboration as they gain access to their research base, while the small firms seek the partnership to access the financial might and distribution channels of the bigger players. Therefore, both firms seek the partnership as a result of their respective resource needs. In reality, there are two ways to illustrate the uniqueness of inter-organizational relationships in the production of groundbreaking resources (Barringer and Harrison, 2000). First, inter-organizational relationships often accumulate a cluster of the best brains (research consortia or multi-firm alliances) to achieve results that no one firm could muster as a stand-alone entity. This was evident when Kodak in the 1990s led an alliance which comprised Canon, Fujifilm, Nikon, Minolta, and others to launch the Advanced Photo System. Harbison and Pekar (1998) insist that this technology was developed simply because of the combined efforts and resources of the individual brands. Secondly, the uniqueness of inter-organizational relationships is creating revolutionary ideas through the amalgamation of firms with adequate market influence. This is evident in the case of Walmart (an American multinational retail corporation).

Headquartered in Arkansas, the corporation has an equity joint venture with Walmex (Cifra), which is the largest division outside the United States and located in Mexico. They develop retail outlets in Mexico that are similar to Walmart outlets in the United States (Harbison and Pekar, 1998). This illustrates a perfect alliance of two firms that have taken advantage of their unique contribution power (market influence, experience, and name recognition) to achieve their aims. The RD theory tends to elucidate the motives behind university–industry collaboration as both parties would consider themselves as resources dependent (Ankrah and Omar, 2015). While the RD theory has a straightforward appeal, it has certain limitations, one of which is the failure to explain the reason why organizations seek other approaches besides alliances to manage resource deficiencies (Child *et al.*, 2005). Also, the theory does not indicate the various ways by which organizational competencies can be developed.

STRATEGIC CHOICE

The strategic choice (SC) theory is a school of thought which explains an organization's strategic decisions to seek alliances for the sake of increasing competitiveness, market power, neutralizing competitors, profit maximization, and increasing market speeds (Barringer and Harrison, 2000; Ankrah and Omar, 2015). This means that decisions are acceptable and tenable if they have a powerful influence on or provide competitive advantage for a firm (Santos and Eisenhardt, 2005). This is evident in the case of the advent of the PowerPC (Performance Optimization with Enhanced RISC – Performance Computing) computer chip which was pioneered by the alliance of Apple, IBM, and Motorola (AIM). Its creation in 1991 was cited as a strategic move by the trio companies (alliance members) to counter the Microsoft–Intel dominance of personal computing (Harbison and Pekar, 1998). The SC paradigm is very broad and motivations from nearly all of the other paradigms can be integrated into this theory. This theory is also relevant as universities and industry may seek collaboration for strategic reasons.

STAKEHOLDER THEORY

The stakeholder theory (ST) opines that stakeholder groups found within an organization play a critical role in the maintenance of social legitimacy (Dacin *et al.*, 2007). This organization's legitimacy can be attained when its activities are aligned with the values and rules of the society in which it operates (Zukin and Dimaggio, 1990). It is worth noting that stakeholders refer to those who directly or indirectly contribute to the firm such as competitors, investors, employees, suppliers, customers, regulatory agencies, and even the local communities (Freeman, 1994). This theory is also relevant as universities and industry may seek collaboration to understand the interests and concerns of relevant stakeholders regarding their strategic decisions and operational expertise (Barringer and Harrison, 2000; Adler and Kwon, 2002).

LEARNING THEORY

The learning theory (LT) emphasizes the role of knowledge in helping organizations gain competitive advantages (Larsson *et al.*, 1998). Considering the tacit and dynamic nature of knowledge, an organization seeking a specific skill stands a better chance of realizing its targets by establishing an alliance with another organization that is knowledge oriented (Barringer and Harrison, 2000). The rationale behind this theory stipulates that firms believe that their competitive position can be further enhanced through superior knowledge (Simonin, 1997). For example, if a firm seeks knowledge in a particular scope or field such as green building or building information modelling, they often stand a chance of learning more by allying with another firm whose knowledge in sustainability is exemplary.

In terms of organizational learning, there are two types of activities that take place during inter-organizational alliances. They are exploitation (improving existing capabilities and reducing costs through increased productivity of capital and firm assets) and exploration (the invention, discovery, or innovation of new opportunities needed for wealth creations) (Cohen and Levinthal, 1990; Barringer and Harrison, 2000). Levinthal and March (1993) also maintain that exploitation and exploration are associated with various time prospects. While exploitation considers current viability, exploration is associated with future viability. Consequently, firms seek inter-organizational relationships as an avenue to share the financial burdens and overheads of exploitation and exploration with their affiliated counterparts (Powell *et al.*, 1996). This perspective therefore argues that UIC can be seen as an approach to generate learning opportunities because collaborations are effective ways for knowledge development and exchange (Hoffmann and Schlosser, 2001). One demerit of this theory is that it focuses heavily on the development and dissemination of skills without taking into cognizance the overheads involved as well as the loss of critical information elements with regard to the collaboration itself (Hamel *et al.*, 1989).

INSTITUTIONAL THEORY

The institutional theory (IT) underlines that organizations are posed with several institutional pressures that compel them to line up with the standards and expectations of their external surroundings (Dimaggio and Powell, 1983). Owing to these institutional exertions and factors, both sets of collaborators (universities and industry) may collaborate to conform to prevailing social norms (Oliver, 1990; Ankrah and Omar, 2015). For example, a firm may seek collaboration for social responsibility, whereas their university counterparts may enter collaborations to appear as accountable, result-oriented, and effective in addressing its challenges.

INTERACTION THEORIES

Unlike the interdependency theory which focuses on the influence of the external environment on collaborations, the interaction theories deal with the internal development of the relationship itself. This theory is chiefly propounded by the social network approach (Brass *et al.*, 2004). Social network studies have been carried out and have dominated several journals in various fields such as management, economics, and marketing amongst others and have explored a wide range of organizational topics across different analytical levels. Network research takes into cognizance what constitutes the general make-up of organizational-related players, including the actors (individuals and work units) and even the organization itself. According to the social network viewpoint, actors are critical components of interconnected relationships that are subject to behavioural discussions and dynamics (Brass *et al.*, 2004). Therefore, a network is defined as a set of nodes and its interconnectedness, while nodes are referred to as actors (individuals and work units within the organizations). The social network perspective also sheds light on the antecedents of networks (actor similarity, personality, proximity and organizational structure, and environmental factors) (Brass *et al.*, 2004).

ACTOR SIMILARITY

Generally, people with similarities tend to interact more easily with one another as this can lead to seamless communication, trust, and reciprocity. However, this is a basic assumption in several theories that have been supported by various researchers (Blau, 1977). Similarity has also been operationally defined on certain constructs such as education, gender, age, social class, prestige, and occupation (Ibarra, 1993; McPherson, 1983). Brass *et al.* (2004) also asserts that similarity is a relational concept. This implies that an individual can be similar with respect to another individual and vice versa and can be dissimilar in relation to others. Factors such as culture and socialization processes can cause an organization to display a certain similarity pattern which Kanter (1977) refers to as homosocial reproduction. This connotes that an individual's similarity pattern in relation to the organization's modal attribute framework may determine the extent to which the individual is integrated into such an interpersonal framework network (Brass *et al.*, 2004).

PERSONALITY

It has become a bone of contention for several researchers who opine that personality is a function of network position. Researchers such as Mehra *et al.* (2001) indicate that personality can indeed affect social network patterns. Their study found that individuals who were central to the studied networks recorded high scores in self-monitoring, which is a stable personality trait. Other researchers such as Klein *et al.* (2004) and Casciaro (1998) argue that personality characteristics can determine friendship types, advice patterns, and perceptions of networks.

PROXIMITY AND ORGANIZATIONAL STRUCTURE

While the focus on similarity and personality suggests that interaction within an organization is simply voluntary, the organizational structure creates a pattern for networks. The structural set-up helps to ensure the division of labour, dissemination of work description (workflow, task design, and hierarchal arrangements), and means for coordination and supervision. The hierarchal flow chart which differentiates various individuals and groups in the organization can restrict interaction opportunities. This is because a line of authority exists which is to be followed, hence networks are influenced by workflow requirements and organizational structures. Burkhardt and Brass (1990) also suggest that new technology can act as a catalyst for varying communication patterns. The advent of recent technologies such as electronic mail has further encouraged some form of cordiality in achieving tasks. For example, actors who are pencilled for a meeting will have to send reminders, hence achieving consensus communication among them (Burkhardt and Brass, 1990; Brass *et al.*, 2004).

ENVIRONMENTAL FACTORS

Several dealings such as mergers, acquisitions, downsizing, and national culture are environmental jolts that have been found to influence network patterns within an organization (Monge and Eisenberg, 1987; Shah, 2000; Brass *et al.*, 2004). Regarding mergers, Danowski and Edison-Swift (1985) observed a significant change in electronic mail usage after a successful merger. However, this scenario lasted for a short time as employees reverted to their previous patterns before the merger (Danowski and Edison-Swift, 1985). In terms of downsizing, Shah (2000) states that firms carry out downsizing in response to several external factors such as technological advances and competitive pressures. This, however, comes with its complexities, hence a successful process is essential for a firm's viability (Shah, 2000). Brockner (1988) suggests that layoffs can elicit several psychological reactions among survivors such as anger and relief which can be connected to motivation, performance, organizational commitment, and job satisfaction.

In summary, the interaction theory assumes the university and industry as autonomous entities and that collaboration or relationship can be achieved by any one of the two taking the initiative. Thus, this theory explores the dynamic of university–industry collaboration and how the alliance evolves through the influence of certain factors such as trust, commitment, and communication.

TRIPLE HELIX OF UNIVERSITY–INDUSTRY–GOVERNMENT RELATIONSHIP

For several decades, developing countries have depended on developed ones for technological assets in a bid to upgrade their industrialization activities. However, this process of technology transfer has had its challenges as the transferred technologies seldom contributed to economic growth and expansion. Three critical reasons for this deficiency have been put forward. Firstly, there is a barrier created in the process of translating the transfer of technology into the development of innovation in developing countries. This is because these developing countries lack adequate capacity to absorb and adapt to acquired technologies. Secondly, most of the acquired technology from developed countries emphasizes the aspect of production capabilities and not the innovation aspect. Finally, most of the technology transfer activities in developing countries align themselves with a linearity between supply and demand sectors which creates a barrier for effective knowledge sharing. Considering these barriers, there is an increasing need for the development of a triple helix (TH) of university–industry-government relations to ensure an innovative and technological process (Saad and Zawdie, 2005; Kharazmi, 2011).

The TH model suggests that universities have a critical and innovative role to play in achieving a knowledge-based society (Etzkowitz, 2008). The model takes into cognizance the relationship between the university, industry and the government and also highlights the internal transformation within each of these individual actors in the relationship in achieving a knowledge-based society. For example, over time, the role of universities has seen it transformed from teaching organizations to teaching and research at the same time. They also play an active role in a nation's community development through their various duties such as entrepreneurial training, educational functions and seminars, amongst others (Etzkowitz and Leydesdorff, 2000). Industry, on the other hand, engages in innovative designs and their transfer, making it a locus of production. Additionally, the model also features the role of government which is to achieve a certain level of balance between intervention and non-intervention to guarantee stable interactions and exchange (Dzisah and Etzkowitz, 2008). In the words of Etzkowitz and Leydesdorff (2000), the TH model focuses on 'the network overlay of communications and expectations that reshape institutional arrangements among universities, industries, and governmental agencies'. One of the main elements of the TH is that any of the three institutions can take over the role of the others. For example, universities can handle business and government roles whilst maintaining their central vision and mission, while industries can perform university functions while still producing goods and services (Etzkowitz, 2008). Furthermore, Hakansson and Snehota (1995) proposed a model to examine and manage the nature of the university–industry-government relations as well as its networks and activities. It is based on this model that the trident is interrelated to enhance the learning and innovation process in a system. The seamless interaction between the trident ensures greater trust and synergy which ensure the success of the TH university–industry-government relations (Saad and Zawdie, 2005).

Employability implications of university–industry collaborations

With the ever-growing demands for graduates to be equipped with more than just an academic degree, several researchers have discussed the influence of UIC on the employability of graduates. From the various theories of UIC examined, it is safe to say that built environment students can benefit immensely from mutual collaborations between both. While the outcomes of UIC for universities and industry (Ankrah and Omar, 2015) were discussed earlier, it is pertinent to understand the possible outcomes on graduate employability. According to Tran (2016), one of the significant roles UIC plays in enhancing graduate employability is ensuring a learning process whereby students can apply acquired classroom knowledge in an authentic work environment. This tends to provide ample exposure for graduates to relevant work contexts, hence UIC enhances experiential learning (Hagan, 2004; Ferns and Moore, 2012) and reflection on acquired knowledge to solve real-life problems (Weisz and Smith, 2010). With experiential learning comes the development of key skills among graduates such as critical thinking skills, teamwork, communication skills, decision-making skills, time management skills, adaptability skills, as well as broader transferable skills (Eames and Cates, 2011; Ferns *et al.*, 2014).

Likewise, the involvement of industry professionals and employers in the university setting and activities such as the delivery of lectures and seminars, involvement in course design, and structuring also bode well for graduate employability (Haupt, 2012, Tran, 2016). Huber and Hutchings (2004) also opine that UIC promotes tacit knowledge development among students. They suggest that students are provided with avenues to master their professional knowledge as well as understand the strengths and weaknesses of their chosen career paths (Huber and Hutchings, 2004; Beckett and Mulcahy, 2006; Zegwaard and Coll, 2011; Ramakrishnan and Yasin, 2011, Elder, 2014). Sternberg *et al.* (2000) highlighted the benefits of tacit knowledge as it helps students to adapt, select, and shape the real world successfully (Sternberg *et al.*, 2000). Artess *et al.* (2011), Helyer *et al.* (2011), Lowden *et al.* (2011), and Rust and Froud (2011) all assert that this type of knowledge has nothing to do with educational achievement or attainment.

Gap 2: Industry 4.0 requirements for graduates

Another gap observed in the reviewed models of employability is the need for built environment graduates to possess certain requirements to fit into the Industry 4.0 movement. With the advent of high-speed networks and smart infrastructures dominating our present-day world, there is a new degree of transformation not experienced since the first industrial revolution.

Introduction to Industry 4.0

Industry 4.0 or the fourth industrial revolution (4IR) is a broad term used to describe the current trend of data exchange, computerization, and automation in manufacturing technologies. Recently, the term has gained popularity that has seen it become a regular topic in the academic field, social groups, as well as industry domain. Several studies have described the 4IR as the blending of technologies of the digital, physical, and even biological world, which leads to the creation of innovative opportunities that influence the economic, political, and social systems. Thanks to new technological achievements which have led to certain terminologies such as artificial intelligence (AI), digitalization and robotization, cloud manufacturing, augmented reality, the Internet of Things (IoT), the Internet of Services (IoS), big data, and cyber-physical systems (CPSs), the 4IR basically transforms modern production (Madakam *et al.*, 2015).

However, the main technology of the 4IR is the CPS which is referred to as the combination of cybernetic (digital) and physical systems (Lee *et al.*, 2015). Both of these systems are interactive in that everything which happens in the physical has bearing on the virtual systems and vice versa in enabling increased efficiency and improvement among firms who adopt this new paradigm (Lee, 2010). The adoption of the previously mentioned technological terminologies is critical to the advent of more intelligent manufacturing processes that comprise devices and production modules that can exchange information, control each other, and trigger actions independently to enable an intelligently working manufacturing environment (Qin *et al.*, 2016). Moreover, the 4IR has attracted varying points of view from firms and even researchers about its conceptual basis and visions. However, there is an apparent consensus about the four main facets and how they address the futuristic vision of the manufacturing sector. These include smart factories, smart products, business models, and the customers (Qin *et al.*, 2016). Furthermore, Industry 4.0 combines and captures manufacturing technologies, contemporary automation, and data exchange. Some of these technologies have resulted in terminologies such as artificial intelligence (AI), additive manufacturing, the Internet of Things (IoT), the Internet of Services, and cyber-physical systems (CPSs) (Madakam *et al.*, 2015). With this new wave of revolution comes a new level of interconnectivity and automation.

SMART FACTORIES

As one of the critical aspects of the 4IR, the future factory is going to become more sensitive and intelligent enough to ascertain the functionalities and maintenance of machines to control and manage the production processes and the factory system, thereby increasing manufacturing efficiency. This new integrative approach will go beyond the conventional paradigm where the manufacturing resources (machines, sensors, conveyors, robots, amongst others) were interconnected for information exchange and project execution. Additionally, several manufacturing processes, including production planning, production engineering, and product design, are not only going to be controlled by a decentralized system but will also be controlled interdependently as a result of a possible modular simulation which connects the processes end-to-end (Lucke *et al.*, 2008; Qin *et al.*, 2016). Simply put, a smart factory environment will not only increase efficiency in the manufacturing sector but will also meet the requirements of complex markets.

SMART PRODUCTS

Another benefit that will arise from 4IR is the emergence of smart products. These technologically driven products will be embedded with various types of sensors, processors, and identifiable components that transmit information and provide a feedback mechanism for customers and other active players (users) involved. Through the use of these elements, several functions can be integrated into these products such as the tracking of products, design optimization, product maintenance, and results analysis (Lucke *et al.*, 2008; Qin *et al.*, 2016). Owing to their ability to establish a functional connection between the digital and physical worlds, smart products can also be referred to as cyber-physical systems. Smart products also possess a high degree of autonomy as well as certain distinct features such as self-identification, data storage, computation, communication, and efficient interaction with their physical environment (Lucke *et al.*, 2008; Qin *et al.*, 2016).

BUSINESS MODELS

The 4IR implies that a holistic communication network is bound to occur between various parties, including the factories, suppliers, customers, and even resources. Hence, business models are impacted by this revolution because there is a new communication pattern along supply chains. In other words, the future of business networks can be enhanced, thus achieving a self-organizing status and transmission of instantaneous responses (Lucke *et al.*, 2008; Qin *et al.*, 2016).

CUSTOMERS

The customers themselves are poised to enjoy certain advantages as a result of the 4IR. According to Qin *et al.*, (2016), customers will be offered a new purchasing method that permits the request of any function whatsoever. Customers, however, will be afforded more opportunities to change their requests at any given time (real-time) and even at the 11th hour without accruing any charges. Identified as a key factor in virtually every business model, customers are assured of improved experience and enhanced communication with the advent of the 4IR. Moreover, smart products will provide customers with relevant information such as product specifics, advice, and other critical details (Lucke *et al.*, 2008; Qin *et al.*, 2016).

The German government first identified this wave of change and technological strategy in 2011, which aims at promoting and improving computerization and industrialization, marking the launching of a 4IR. The nation has also earmarked it as a strategic approach to further develop its economy in the next few decades (Kagermann *et al.*, 2013). Several other nations in Europe and even the Asian continent are gradually embracing the theme (Zhang *et al.*, 2014). Because the 4IR is still relatively new, its implications and consequences are still considered as projections, assumptions, or forecasts. Several benefits such as the production of cheaper products and the efficient usage of resources, as well as barriers such as income inequalities and unemployment possibilities, are widely argued. Nevertheless, it may possibly take three or four decades to experience these possible benefits and barriers of this new revolution. But what exactly makes Industry 4.0 a revolution? Also, what are the previous industrial revolutions, and how can they be compared to the 4IR?

FIRST INDUSTRIAL REVOLUTION (1750–1850)

This era was characterized by the adoption of water, coal, and steam power to execute human activities towards the end of the 18th century. The emergence of the first industrial revolution can be traced to Britain and its arrival was marked by an efficiency jump in a few sectors, starting with the material industry before moving to various enterprises later on (Freeman and Soete, 1997). The principles started with the Watt steam motor and water wheel which brought about several developments as a result of its applications, paving way for the emergence of the water and steam engine (Brynjolfsson and McAfee, 2014). This revolution resulted in radical changes in both social and financial life cycles and brought about the division of work, development of urban focuses, and the rise of a lowly class, amongst others (Hobsbawm, 2016). For this era, in terms of employment and economic value, the dominant industry was the textiles industry. This industry was also noted as the foremost to adopt modern production methods. Other predominant technologies that were peculiar to the first industrial revolution include iron making and steam power.

SECOND INDUSTRIAL REVOLUTION (1850–1914)

As the first industrial revolution wound down towards the end of the 19th century, the second industrial revolution was heralded, this time by the United States (Freeman and Soete, 1997). This era was largely characterized by the emergence of chemical and electrical energy (electric power) replacing the steam engine. This had significant impacts on enterprises such as steel, railways, and synthetics, amongst others (industrial mass production). It was during this era that the famous Eiffel Tower, one of the most visited tourist attractions in the world, was built.

THIRD INDUSTRIAL REVOLUTION (1969–PRESENT DAY)

As in the case of the second industrial revolution, the United States was the frontrunners of this revolution. However, the Asians became a critical player as well (Freeman and Soete, 1997). The critical components of this revolution were propounded by the advent of personal computers, the Internet web, and integrated circuits (microchips) which were brought about by research and development by governments and educational institutions (Freeman and Soete, 1997). Even in the manufacturing sector, hardware and information technology (IT) provided a robotic experience for executing tasks that had previously been executed by physical means. With the advent of technological influence, several terms, including advanced manufacturing technologies (AMT), prevailed in the 1980s which led to the introduction of several technologies, including PC-helped outline (CAD), PC-incorporated manufacturing (CIM), adaptable manufacturing frameworks (FMSs), and PC-supported manufacturing (CAM), amongst others (Gerwin and Tarondeau, 1982; Lei et al., 1996). These various technological processes helped foster adaptability, improved automation, the exactness of process, improved control, and shorter creation cycles (Goldhar and Jelinek, 1983). Therefore, the third industrial revolution can also be regarded as the digital age which spanned from the 1970s to the early part of the 21st century.

The era further saw the arrival of electrical devices and microelectronics, which led to technologies such as the World Wide Web (www), cell phones, and tablets, amongst others (Du Plessis, 2017).

Hence, in their respective orders, the first three industrial revolutions were as a result of mechanization, electricity, and information technology. However, the arrival of IoT and IoS into the manufacturing setting ushers in a new dimension known as the 4IR. Therefore, the core of every industrial revolution brought about the need for increased efficiency and productivity. As already discussed, the first three industrial revolutions brought about increased efficiency and productivity in industrial processes through the adoption of random technological developments, from the steam engine to electricity and then digital technology. However, the 4IR represents a more complex technological paradigm which has generated several discussions about its influence in the industrial sector.

Features of Industry 4.0

Unlike the previous revolutions, the 4IR is still being predicted and gradually implemented, which allows governments, researchers, companies, and other stakeholders to take the necessary actions in ensuring adequate preparedness for this wave of transformation. Oesterreich and Teuteberg (2016) insist that from a technical perspective, this new paradigm leads to increased automation and digitization which is enabled by the creation of a digital value chain. The main features of the 4IR concept can be further explained through four dimensions of integration: horizontal integration through value networks, vertical integration, networked manufacturing systems, and end-to-end digital integration of engineering across the entire value chain. The horizontal aspect deals with the integration of several processes, information, and technology systems and information flows within a firm and between other firms. In the case of the vertical aspect, the integration of systems through the various hierarchical levels of firms is the biggest concern. The major objective of both the horizontal and vertical integrations is to achieve an end-to-end result across the whole value chain through the use of CPS (Henning *et al.*, 2013). Hermann *et al.* (2016) also suggest that the 4IR can be described as an all-encompassing term for concepts and technologies which encapsulate the entire value chain of organizations.

However, there are some striking observation and components which will be discussed in detail. Firstly, manual labourers (human workforce) within the factory are conspicuously absent, as most of the workflow activities within the manufacturing process have been automated. Secondly, production information is transmitted via interconnected machine sensors, producing big data that are stored on a decentralized cloud that is used to produce several performance indicators in real time. Thirdly, 3D printing is introduced, which allows users to explore high levels of customization features such as the Internet. Fourthly, there are horizontal red and black arrows, which depict the complete incorporation of the key stakeholders, including the suppliers, manufacturers, and end customers. The final observation refers to the resources of the future, a feature that falls outside the project scope. This aspect includes renewable energy sources that are focused on the concept of sustainability (Henning *et al.*, 2013; Hermann *et al.*, 2016; Du Plessis, 2017).

Critical components of Industry 4.0

The 4IR is a very fast and complex technological system that is underpinned by connectivity and production digitization which emphasizes the need for the integration of all elements in a value-adding system and industrial processes (Pereira and Romero, 2017). Likewise, the concept also embraces digital manufacturing technology, automation, and computer technology as well as network communication technology. The hallmark of this revolution is the advent of technological advancements for eradicating the borders between the digital and physical worlds and integrating human and machine agents, products, materials, and production systems (Henning *et al.*, 2013). The next section describes the critical technological drivers (central underpinning aspects) of the 4IR.

CYBER-PHYSICAL SYSTEMS (CPSS)

The CPSs which are regularly used to describe the 4IR signify one of the most noteworthy developments regarding information technologies and computer science advances. Kagermann *et al.* (2013) suggest that CPS is a pivotal component of the 4IR as it is the synthesis of both the physical world and the virtual world. Broadly speaking, CPS can be described as innovative and transformative technologies that support the management of system interconnectedness through the incorporation of their computational and physical environments (Lee *et al.*, 2015). The adoption of the 4IR in the manufacturing sector requires an industrial implementation of CPS, also referred to as cyber-physical production systems (CPPSs), which, when applied to the production process, will play a crucial role in the connection systems ranging from autonomous elements to the cooperative ones. Similarly, a CPS can be referred to as an embedded system that encourages data exchange in an intelligent network, bringing about smart production (Henning *et al.*, 2013). Also, when CPS is connected to the Internet, it is commonly regarded as the Internet of Things (Pereira and Romero, 2017).

In the study by Hermann *et al.* (2016), three developmental stages of CPS were identified. The first developmental stage deals with the aspect of CPS that focuses on unique identification which is possible through identification technologies. In the second stage, various entities within the CPS domain are furnished with sensors that allow data capturing possibilities. For the third phase, the CPS consists of several entities with numerous sensors that possess data-analysing abilities. This three-stage definition of CPS is also supported by Du Plessis (2017) who highlights CPSs as real-time (physical) systems which are supported by partial intelligence and can achieve the following: make decisions via sensors, use actuators to impact the environment and system, store data, analyse data, and establish interactivity with the digital and physical world.

The IoT is a term that combines several technological approaches based on the interconnection between the Internet and physical things. Since the arrival of the Internet, the interconnectivity between computer systems has waxed stronger, and the technological strides over the years have paved the way for the Internet to evolve to the next level, resulting in smart(er) objects (Henning *et al.*, 2013; Pereira and Romero, 2017). Hence, the smart object is a spin-off from the IoT, because this new revolution is characterized by the interaction and interconnection of intelligent objects with their environment and other objects. Therefore, IoT can simply be described as the Internet connection between things (machines and robots), people, systems, and objects (smartphones, tablets, and laptops), all interconnecting to create a smart environment-cum-smart factory to achieve definite goals (Henning *et al.*, 2013; Hermann *et al.*, 2016). It is also defined as a world where physical objects are incorporated into the information framework or network and can become active participants in the work processes (Pereira and Romero, 2017). Sundmaeker *et al.* (2010) suggest that the IoT offers both people and things the opportunity to stay connected at any point in time, anywhere, with anything as well as anyone while using any network with any service provider (Du Plessis, 2017). In various industrial set-ups and value chains, the IoT will offer several opportunities for companies, manufacturers, and users to have a significant impact in various fields such as process management, logistics, industrial manufacturing, automation, and transportation (Henning *et al.*, 2013). Subsequently, in describing the impact of IoT in the industrial sector, the term 'Industrial Internet of Things' (IIoT) has been coined, as it deals with the adoption of disruptive technology such as control systems, data analytics, actuators, security mechanisms, and sensors to advance the current industrial systems. The constant spread of IoT techniques will ensure things become more reliable, more autonomous as well as smarter, paving way for added-value products and services.

This recent concept will provide fresher opportunities to the service industry because it deals with business network creation between the customers and the service providers (Pereira and Romero, 2017). This concept has similarities to the IoT in its approach, but it deals with services as against physical entities. The IoS can be regarded as a new business model that will create a new approach for service delivery, paving the way for a higher value creation which comes as a result of interconnection among the stakeholders of the value chain, including the customers, suppliers, intermediaries, and organizations as a whole. The goal of the IoS is to utilize the Internet in a variety of innovative ways to create value for the services sectors (Terzidis *et al.*, 2012), as well as ensure that services are maximized in a digital medium (Du Plessis, 2017). In addition, the concept of service, as seen from the information technological perspective (IT), refers to a technical understanding of the various functions of software that are put forward as web services. Plass (2015) also describes the IoS as the expansion of communication and networking links within the IoT, which in turn generates big quantities of data that can be accessed via the cloud. These data are known as big data.

A smart factory can be described as a factory that is responsible for supporting both machines and human labour in executing set-out activities (Hermann *et al.*, 2016; Kagermann *et al.*, 2013). This is achievable by certain passive background systems which are known as calm systems. These systems can understand the information context of an entity, hence their name. These systems tend to complete their tasks by absorbing information from both the digital and physical world and the synergy of both worlds is regarded as a CPS (Kagermann *et al.*, 2013). Furthermore, it is also significant to note that a smart factory comprises a CPS through which a calm system, including hardware and software, can relay messages between humans and machines (Du Plessis, 2017). Hence, the smart factory is one of the components of the 4IR.

Building blocks of Industry 4.0

Often known as the pillars of the 4IR, the building blocks can be regarded as the various supporting technological tools which are pivotal to the advancement of the 4IR. These technological tools are in existence, but require some necessary modifications so that they can be user-specific (Du Plessis, 2017). The integration of these technologies will herald the arrival of Industry 4.0. These technologies include additive manufacturing, augmented reality, big data analytics, communication technology, cloud computing, cybersecurity, digital factory, embedded electronics, and simulation.

ADDITIVE MANUFACTURING

One of the ways by which manufacturers can improve the level of customer satisfaction is through product customization (Rüssmann *et al.*, 2015). Thankfully, this can be achieved by additive manufacturing as it provides manufacturers with a variety of innovative ways to offer improved value to end-users whilst saving cost and time for the manufacturer. One of the processes by which establishments have begun the adoption of additive manufacturing is the use of 3D printing which is mostly utilized for the production of distinctive components as well as prototyping (Rüssmann *et al.*, 2015). The adoption of additive manufacturing methods will help in the production of customized products which will offer several benefits in terms of design make-up and functionality (Du Plessis, 2017).

AUGMENTED REALITY

Although still in their infant stage, augmented reality-driven systems are poised to have a significant effect on the support systems of establishments in the near future (Rüssmann *et al.*, 2015). It is expected that establishments will adopt these technologies to provide real-time information for their workers, which will help in improved decision-making and project realization. This can be achieved in the form of virtual training for both old and new workers, thereby equipping employees to handle dynamic situations.

BIG DATA ANALYTICS

Establishments and businesses cannot ignore the concept of data as it is critical to the entire production process. The term 'big data' refers to datasets, the magnitude of which goes beyond the ability of well-known database software tools to record, keep, run, and analyse (Manyika *et al.*, 2011). Big data can be differentiated from traditional data in three significant aspects: volume, velocity, and variety. Volume deals with the size of the datasets, velocity deals with the rate of data generation, and variety deals with the various categories of data collected (Manyika *et al.*, 2011). Through predictive maintenance, big data analytics and its technologies can lead to quality products, improved decision-making, manufacturing flexibility, and energy efficiency by supporting instantaneous data collection from different sources as well as a comprehensive analysis of data collected (Bahrin *et al.*, 2016). In the 4IR, there are six C's that underpin big data and big data analytics (Hilbert, 2013). These include cloud computing, content and context, community data sharing and collaborative partnerships among key stakeholders, the connection between networks and sensors, cyber which deals with the system properties, and customization which deals with the translation of data to achieve results (Hilbert, 2013).

COMMUNICATION TECHNOLOGY

The proliferation of communication technology is simply staggering in that during this decade, over 60 billion devices will have an established connection with Internet services (Evans, 2011). This surge places a high demand on the number of Internet Protocol (IP) addresses that will be required as well as a more functional and reliable communication protocol. This has led to the advent of Internet Protocol version six (IPv6) which dealt with the problems of IPv4 address exhaustion. IPv6 thus provides several technical benefits such as the permission of a larger number of devices to establish connections around the world (Du Plessis, 2017).

CLOUD COMPUTING

The 4IR and cloud manufacturing support the networking and interconnection of data, things, services, and people over the Internet in the manufacturing world, encouraging a smart system. Cloud technologies ensure increased data sharing across company boundaries, leading to improved system agility, productivity, and flexibility whilst reducing cost implications (Liu and Xu, 2017). Cloud computing can further be described as a model for permitting a wide range of on-demand network access to an interconnected pool of computing resources, including servers, services, and networks (Mell and Grance, 2011). While Armbrust *et al.* (2010) suggest that cloud computing services require minimal service provider interaction, Du Plessis (2017) states that it can also be regarded as software as a service (SaaS), with reference to the operations of Google Drive. In the case of Google Drive, which is a synchronization service, there is not only room for a storage service but also other services such as Google Docs, Sheets, and Slides. The ubiquitous nature of 4IR ensures that large datasets are involved which guarantees that data sharing will become even more prominent. Moreover, cloud technologies have become more developed for various production systems owing to the constant efforts of developers.

CYBERSECURITY

With several establishments and businesses discovering the benefits of going online, high demand for higher cybersecurity is inevitable (Sundmaeker *et al.*, 2010). Considering the quantity and nature of data that are shared in networks, various industrial networks and systems are more prone to cyber threats than ever before. To mitigate this, cybersecurity measures and processes have to be installed to prevent damage and theft, amongst other cyber threats (Gasser, 1988; Du Plessis, 2017).

DIGITAL FACTORY

According to Bouri *et al.* (2011), the digital factory is a connecting bridge between a virtual and a real factory. The digital factory is also a real-time recreation of the real factory, which can be used to carry out vastly accurate simulations that are under investigation (Gregor *et al.*, 2009). Kuhn (2006) refers to a digital factory as a broad term that encompasses the conjunction of digital network tools and models such as simulation and 3D visualization. This kind of system can be beneficial by improving the planning, processes, and resources of the real factory. The core objectives of the digital factory include communication improvement, data management, improving planning quality, industrial improvement, and standardization. According to Bracht and Masurat (2005), both communication and data management can be enhanced with cloud computing, as well as master execution systems (MES) and enterprise resource planning (ERP). Both the industrial improvement and planning quality can be enhanced by avoiding negative investment and a well-planned market introduction of new products.

EMBEDDED ELECTRONICS

Embedded electronics in this context refer to microchip technologies, the key functionalities of which have been applied in various capacities (hardware and software) within certain system constraints. Hence, embedded electronics highlight the use of robots, making it a critical component of the 4IR.

Present-day robots are systems that offer flexibility, cost advantage, autonomy, and cooperation. Kamble *et al.* (2018) predict that robots may soon start the process of interacting with one another as well as establish a rapport with humans in solving problems. According to Sundmaeker *et al.* (2010), another form of electronics that connects with IoT and the 4IR are the open-source programmable logic controllers (PLCs) which are loaded with several add-ons and modules that ensure their connection to the Internet and the CPS of an establishment.

SIMULATION

According to Du Plessis (2017), simulation has provided establishments with the opportunity to test alternatives, simulate processes, and carry out feasibility studies of projects. This invariably reduces the cost of trial and error for establishments. According to Kamble *et al.* (2018), both simulations and prototyping will rely on instantaneous data to create a mirror representation of the physical world in a virtual model, which can include humans, products, and even machines. Simulation can further observe machine behaviours and operational connectivity, prevent deadlocks, and carry out real-time tracking of manufacturing cost and the energy efficiency of wireless sensor networks, amongst other functions (Lin *et al.*, 2016; Moreno *et al.*, 2017; Ramadan *et al.*, 2017).

Design principles for Industry 4.0

According to Hermann *et al.* (2016), seven basic principles underpin the concept of 4IR: decentralization, interoperability, modularity, real-time capability, service orientation, standardization, and virtualization.

DECENTRALIZATION

Decentralization is defined as the ability of CPSs within smart factories (companies, staff, and machines) to create and implement decisions solely on their own, instead of depending on a centralized system for decision-making. This is because a centralized control system results in a slower decision-making system as well as a limited ability to customize individualized products (Hermann *et al.*, 2016). Decision-making in this context is not a system of artificial intelligence (AI) but rather a system of embedded electronics (Du Plessis, 2017). Moreover, this principle offers flexibility and helps in quick decision-making to achieve results (Sackey *et al.*, 2017).

INTEROPERABILITY

According to Du Plessis (2017), interoperability refers to the ability of all entities (humans, machines, and products) within a manufacturing environment to communicate seamlessly with each other. In other words, it means that every 'thing' and 'object' that constitutes the CPS has certain communication protocols to interpret data coming to and from each other. According to Qin *et al.* (2016), interoperability provides a trustworthy system whereby multiple networks within the manufacturing system can be extended.

MODULARITY

According to Hermann *et al.* (2016), modularity refers to the capability of systems to be flexible and adaptable to significant changes and external inputs by the replacement or advancement of individual modules. Qin *et al.* (2016) suggest that modularity also facilitates the simulation of several manufacturing practices and processes such as production engineering and planning, product design, and production services as interconnections between them provide a certain level of interchangeability (Du Plessis, 2017).

REAL-TIME CAPABILITY

Real-time capability is the propensity for a system to identify, capture, extract, and analyse data almost immediately. For establishments, this implies that data collection and analysis happen in real time, which improves efficiency (Hermann *et al.*, 2016). For example, in the case of machine failure, tasks are rerouted almost immediately to other machines with the necessary changes and schedules updated automatically. This ability helps in maintaining the decision-making speed and profitability of Industry 4.0–based firms (Sackey *et al.*, 2017).

SERVICE ORIENTATION

The service orientation element of the 4IR is the integration of the IoS in the working space of an establishment, leading to the creation of a product-service system. It is further described as the ability of 'things' and 'objects' to effectively connect with and have access to CPS services and the smart factory through the IoS (Hermann *et al.*, 2016). The outcomes of the service orientation (agility and flexibility) help establishments to be more adaptable to market dynamics more quickly. This paves the way for the collaboration and coexistence of organizational stakeholders to co-create value for end-users (Sackey *et al.*, 2017).

STANDARDIZATION

Plass (2015) motivates its inclusion as part of the design considerations for Industry 4.0. This stems from the need to manage the level of standardization and collaborative mechanisms of information exchange. The need to standardize variables, including channels, communication protocols, frameworks, and data standards was highlighted severally in chapters of the book titled *Vision and Challenges for Realising the Internet of Things* by Sundmaeker *et al.* (2010).

VIRTUALIZATION

Virtualization simply involves the creation of a virtual copy of the physical world and it is used for machine-to-machine communication and process monitoring. According to Hermann *et al.* (2016), it is the propensity for a system to monitor physical processes through simulation software and live sensor feed. These mediums provide any manager with real-time updates as well as an overview of the entire system on a smart device (Sackey *et al.*, 2017).

The place of learning factories in Industry 4.0

The 4IR movement means that there will be a shift of skills required to support these colossal industry changes. Likewise, the complexities of construction 4.0 imply that higher education will have to work in tandem with industry to ensure built environment graduates are adequately skilled to handle these next-generation challenges. Moreover, traditional lectures and teaching methods have been found wanting and have shown limited effects in preparing the younger generation for the complexities of the 4IR. This increases the need for new learning approaches that provide the following: adequate and realistic training of young engineers, modernization of the pedagogical experience, adoption of new manufacturing knowledge during industrial practice, and the improvement of capabilities among trainees such as creativity, problem-solving capabilities, and system-thinking abilities, amongst others (Abele *et al.*, 2015). This has led to the institution of several learning factories around the world in a bid to bridge the gap between formal education and real-world experiences to ensure a holistic educational experience for built environment graduates. In recent times, they have become an effective knowledge and competence platform that provides training, education, and research within both the academia and industry (Sackey *et al.*, 2017). The objectives of learning factories vary from organizational and technological (research-based) to developmental (educational training) in helping users/trainees understand a complex and very new environment.

The term 'learning factory' came into existence in 1994 by the National Science Foundation (NSF) in the United States. Through a consortium pioneered by Penn State University, the foundation was granted the license to institute a learning factory that dealt with interdisciplinary practical senior engineering design projects related to the industry. In 2006, the programme received the National Academy of Engineering's Gordon Prize for Innovation in Engineering Education. The early idea of learning factories emphasizes the practical experience acquired by the application of knowledge learnt during construction education in solving real-life industry problems (Abele *et al.*, 2015). In recent times, the adoption of learning factories has become predominant in parts of Europe and its facilities have varied in sizes and complexity, aiming to improve the learning process of trainees in various areas of expertise.

According to Wagner *et al.* (2012), a learning factory consists of two words: 'learning' and 'factory'. Learning emphasizes the benefits of experiential learning

(learning by doing or hands-on experience) which paves the way for a greater level of retention as opposed to conventional approaches such as lectures (Cachay *et al.*, 2012). In learning factories, trainees are provided with opportunities to test approaches, discover facts, and conduct experiments on several organizational and technological-related issues. Hence, the primary goals of learning factories include organizational or technological innovation, or both (research-based), competency development (training and education–based), and overall employability (including emotional and problem-solving). Furthermore, Abele *et al.* (2015) suggest two significant learning factory classifications. These include the learning factory in the wider sense that is less practical (virtual reality) and can address a wide variety of problems and the learning factory in the narrow sense (physical-based learning factory) which offers a real value chain for a physical product on-site.

The evolution of Industry 4.0 around the world

After understanding the basic elements of the various phases of the industrial revolution (first, second, third, and fourth), it is pertinent to know how various regions and countries have received the latest concept (4IR). Since the term was first coined in 2011, coupled with the increased attention around IoT and CPS, several governments, sectors, industry, and society in general, have observed the sudden tilt towards the 4IR and have begun to draw up policies and strategies to benefit from the advent of this new movement (Siemieniuch and Sinclair, 2015). Liao *et al.* (2017) also observed the increase in the number of research studies (conferences and scholarly materials) that addressed the 4IR concept, suggesting a significant increase between 2013 and 2015 with the number doubling within that span. The growing interest around this topic begs the question: how have various countries and regions embraced the 4IR concept? The next section provides a summary of the various public policies of various countries and regions regarding the 4IR.

THE CONCEPT OF INDUSTRY 4.0 IN ASIA

Initially proposed in 2014, the Chinese Academy of Engineering recommended the Made in China 2025 concept. The Chinese Ministry of Industry and Information Technology (MIIT) was subsequently tasked with the construction and outlining of the subsequent strategic plans. A year later, the Chinese State Council (SC) formally approved the Made in China 2025 approach and it became a national action programme besides the Internet Plus. The plan prioritizes the needs of ten fields in the manufacturing sector in a bid to accelerate industrialization in China (Liao *et al.*, 2017; Liao *et al.*, 2018).

The concept of 4IR in Japan dates back to July 2010, according to the book by Dr. Hiroshi Fujiwara (2010). This book advocates the adoption of electric cars and solar energy in a smarter way to bridge the gap between consumption and production. In 2015, the Japanese Cabinet Office, through its Council for

Science, Technology and Innovation (CSTI), published the 5th Science and Technology Basic Plan aimed at realizing a super-smart society (Liao *et al.*, 2017, 2018).

In 2014, the Indian Prime Minister, Shri Narendra Modi, announced the formation of the initiative called Make in India. Managed by the Department of Industrial Policy and Promotion (DIPP), this initiative was designed to transform India into a global design and manufacturing hub by encouraging establishments to produce their products in India. This policy was distinct from the others as it set out to develop 25 designated sectors of the Indian economy (Liao *et al.*, 2017, 2018).

Meanwhile, in Malaysia, the manufacturing industry is one of the seven key areas the nation has identified over time. Through the Eleventh Malaysia Plan constituted by the Economic Planning Unit (EPU) in 2015, the nation has made considerable efforts in transforming its three main subsectors, namely chemicals, electrical (electronics), and the machinery, towards enhancing their value and diversity (Liao *et al.*, 2018).

Published by Singapore's National Research Foundation (NRF), the Research, Innovation and Enterprise (RIE 2020) national strategy is aimed at improving the research of industries and nurturing innovative enterprises to meet its national needs (Liao *et al.*, 2018).

In 2014, South Korea's Ministry of Trade, Industry and Energy (MOTIE) constituted the Manufacturing Innovation 3.0 in a bid to keep up with the innovative trends of modern manufacturing (South Korea, 2014). This initiative is aimed at the promotion and enhancement of thousands of factories by the incorporation of more advanced systems such as 3D printing by the year 2020 (Liao *et al.*, 2017, 2018).

Finally, the Taiwan Productivity 4.0 Initiative was instituted in 2015 as a means of responding to the advent of the 4IR in Taiwan. This initiative promotes the integration of smart technologies to upgrade and transform not only the manufacturing industries but also the service and agriculture industries (Liao *et al.*, 2017, 2018).

THE CONCEPT OF INDUSTRY 4.0 IN EUROPE

As in the case of Asia, the European continent has instituted several initiatives to meet up with the 4IR. In 2013, the European Union's Factories of the Future (FoF) was made known under the European Union Framework Programme for Research and Innovation known as Horizon 2020. This initiative planned to make available almost 1.15 billion euros for several FoF activities within a six-year span between 2014 and 2020. Recently, the Factories 4.0 and Beyond initiative was established by the European Factories of the Future Research Association (EFFRA) to offer necessary updates with regard to the wave of the 4IR.

In 2013, the French government established the La Nouvelle France Industrielle, which defines 34-sector-based initiatives as the most important in the nation's industrial sector. In 2015, the second phase of this initiative was executed through the modernization of the nation's production tools and the offering of nine industrial solutions which address major socio-economic challenges.

In Germany, the term 'Industry 4.0' made its first appearance in 2011 and attracted more attention when it became one of the ten cardinal projects within the High-Tech Strategy 2020 initiative in 2012 (Liao *et al.*, 2017). This action plan is aimed at developing cutting-edge technologies in a bid to further fortify and secure the future of the German construction industry.

The Fabbrica Intelligente, an Italian technological cluster set up by the Italian Ministry of Education, Universities and Research (MIUR) to support and promote the activities of the nation's manufacturing industry, has existed since 2012. In 2016, the Italian Ministry of Economic Development formed the Piano Nazionale Industria 4.0, an initiative to keep up with the waves of the 4IR and should be implemented between 2017 and 2020.

In 2015, the Spanish Ministry of Industry, Energy and Tourism (MINETUR) established the Industria Conectada 4.0 to enhance the digital transformation of the Spanish industry through its private and public stakeholders (Spain, 2014b).

In 2013, Teknikföretagen (the Association of Swedish Engineering Industries) proposed the Made in Sweden 2030 research initiative. This action plan identified six critical areas to enhance the production and implementation of research and innovation. Three years later in 2016, the Swedish Ministry of Enterprise and Innovation instituted the Smart Industry strategy that aims at addressing four focus areas, with Industry 4.0 being one of them.

Finally, the Future of Manufacturing action plan was established in 2011 in the United Kingdom in a bid to provide long-term planning on ways the manufacturing sector can be improved until 2050.

THE CONCEPT OF INDUSTRY 4.0 IN NORTH AMERICA

The Canadian Manufacturers and Exporters (CME) developed the Industrie 2030 initiative alongside the Canadian Ministry of Innovation, Science and Economic Development to improve the value added by the manufacturing and service industries by 2030. In 2016, the initial phase of this initiative was completed and the reports were issued during the nation's National Manufacturing summit (Liao *et al.*, 2017, 2018).

In 2016, Mexico's Ministry of Economy developed the Crafting the Future initiative to provide a long-term vision and action plan for the future of its manufacturing sector. Over time, the nation has become an excellent manufacturing hub because of its competitive advantage with regard to its high-volume production and relatively low-cost labour (Liao *et al.*, 2017, 2018).

In the United States, in 2011, the President's Council of Advisors on Science and Technology (PCAST) proposed the Advanced Manufacturing Partnership (AMP). This action plan aimed to establish an alliance between academia, industries, and the government to encourage innovation and fresher ideas for design technology and methodologies. In 2013, an improved version of AMP was recommended which was known as AMP 2.0. This upgraded version was launched in a bid to enable innovation as well as improve the business climate.

Industry 4.0 implications

Based on the various impact of Industry 4.0 globally, the concept has created several opportunities for several nations. However, owing to ongoing automation and digitization, several contemporary challenges have risen (Hecklau *et al.*, 2016). These include economic, political, social, and technological challenges, which are discussed in the following sections.

ECONOMIC CHALLENGES

Owing to globalization, present-day businesses and companies are tasked with the responsibility of dealing with shorter product life cycles, cost-cutting, and reduced time-to-market to remain competitive. Likewise, Hecklau *et al.* (2016) argue that the modern-day market has become increasingly heterogeneous and volatile. Owing to the various features and components of Industry 4.0, this era heralds the advent of entirely new industries, reshapes the boundaries of various business sectors, and establishes new competitive challenges among innovative firms. Hence, to stay competitive, companies and businesses will be required to establish strategic alliances with competitors and service providers. Consequently, this move is expected to increase the complexity of processes, leading to financial constrictions (Luthra and Mangla, 2018).

POLITICAL CHALLENGES

It is expected that the era of Industry 4.0 will stimulate governments around the world to support companies and businesses by providing and developing cutting-edge technologies and integrating them into society. With the big data features of the 4IR, governments are posed with the daunting task of establishing legal restrictions to address and manage data collection and privacy protection while dealing with smart things and smart objects (Hecklau *et al.*, 2016; Luthra and Mangla, 2018).

SOCIAL CHALLENGES

The advent of the Industry 4.0 era is poised to deliver social challenges ranging from employee skill adequacies to acceptance of cutting-edge technologies. Owing to the ongoing automation and digitization, the level of digital culture and skill requirements calls for concern. Companies and businesses are expected to embark on retraining graduates (employees) to work comfortably in a dynamic ecosystem environment. This is because companies' success in this innovative era will depend heavily on their employees' knowledge, skills, and competencies (Geissbauer *et al.*, 2016). With jobs becoming increasingly automated and digitized (using software and analytics), it will be necessary for employees to be fully qualified to make the necessary decisions and ensure output generation. Hence, there will be an exponential increase in jobs requiring advanced skills (qualifications) and a decline in jobs requiring lower skills (qualifications) (Hecklau *et al.*, 2016).

TECHNOLOGICAL CHALLENGES

Apart from economic, political, and social challenges, the advent of the Industry 4.0 era is poised to deliver technological challenges that result from digital connectivity. This has created several high-tech challenges for companies and establishments such as cyberattacks (industrial spying) and data rights acquisition issues (Zhou *et al.*, 2017). It is recommended that companies and businesses endeavour to innovate their current business models (Brettel *et al.*, 2014).

Theory backing the 4IR concept

While the 4IR is a relatively new and developing concept, it is essential to discuss how this innovative line of thinking is backed by foundational ideologies. While theories governing this wave of digitalization are still being developed, this study leans towards the theory developed by Klaus Schwab. According to Schwab (2017), the 4IR is 'characterized by a fusion of technologies that is blurring the lines between the physical, digital, and biological spheres'. Based on Schwab's theory of industrial revolution, several technologies will be the driving force of fundamental changes. These technologies include robotics, 3D printing, artificial intelligence, nanotechnology, materials science, data analytics, and several other digital components (Schwab, 2017). These technologies are certain to play a significant role in changing the anthropological practices of present-day individuals and their activities. The proposed theory by Schwab postulates that the 4IR signifies a transition in which the classic metaphor Cogito ergo sum ('I think, therefore I am') attains a different dimension. The modern era is largely based on man's understanding that as a proponent of consciousness, his activities play a big part in defining his relationship with other people and activities carried out.

Therefore, the 4IR era provides an opportunity for the manifestation of artificial intelligence technologies as they work closely with human intelligence, otherwise called cyberphysical systems. This interrelationship between artificial intelligence and human intelligence is twofold. Firstly, traditional human functions and daily activities have begun to be operated by artificial intelligence systems, which ensure a major revision of human anthropological practices. Secondly, the possibility exists of a loss of jobs (social component) due to the newer demands of the workplace (Kravchenko and Kyzymenko, 2019). This was echoed during the 2015 World Summit on Technological Unemployment held in New York, USA that:

> accelerating technological employment will likely be one of the most challenging social issues in the 21st Century. Never before in history are so many industries being simultaneously upended by new technologies. Though 'creative destruction', in which lost jobs are replaced with new ones will be a factor, our newest technologies have the clear potential to eliminate many more jobs than we create. With advanced technology at a geometric pace, robotics, artificial intelligence, 3D printing, and other innovations with enormous disruptive potential will soon hit the main-stream. Billions of people worldwide are currently employed in industries that will likely be affected – and billions of new entrants to the workforce will need jobs.

This book focuses on the two-pronged aspects of Schwab's theory and proposes that, owing to the significant changes that accompany this recent wave of digi-talization (4IR era), a paradigm shift in education is required to develop students for the future. This can be achieved by integrating 4IR components into the already existing curricula to develop students who can handle the ever-increasing changes brought about by the 4IR. Hence, this book is valuable, relevant, and timely as the employability discussion gathers pace.

Skills for the future

To enjoy the dividends of the 4IR, the training and qualifications of built envi-ronment graduates will have to be shaped to align with the new cutting-edge innovations in the digital space. These graduates will require a plethora of new and improved skills sets as automation continues to advance. Hence, universities are faced with the daunting task of embracing strategic plans that will intro-duce digital technology into conventional lecture activities and ensure a more active learning approach to stimulate the interest of learners (Gleason, 2018). One of the necessary skills required for the 4IR era is lifelong learning (con-tinuous learning), which implies the willingness and capability of learning inno-vative concepts anywhere and anytime. Individuals who possess the attitude to continuously learn, display curiosity, show willingness to change their thinking, and display flexibility when the need arises will be highly sought after. With the dynamism of the technologies embedded in the 4IR, individuals will be required to show curiosity and willingness to learn new ideas (Xing and Marwala, 2017; Motyl *et al.*, 2017). Continuous learning also provides an opportunity for students to develop their communication skills as they interact with various professionals from various 4IR domains. In this innovative era driven by the 4IR, built envi-ronment graduates are expected to be digital savvy. They will be expected to have broader perspectives of ideas and innovations (systems thinking), understand and implement business opportunities (business thinking or resource management skills), understand the various dimensions of technological innovation (data mining and robotic manipulation), and display soft skills and problem-solving skills (Aulbur and Bigghe, 2016; Gray, 2016; Xing and Marwala, 2017; Motyl *et al.*, 2017).

Furthermore, the adoption of software, applications, analytics, and connectivity is prevalent in the 4IR era, and this will increase the demand for graduates with sense-making skills (ability to determine the deeper meaning of innovations), computational thinking (ability to understand data-based reasoning), new-media literacy (ability to assess and develop content in advanced media forms), multidisciplinarity (ability to understand concepts across numerous disciplines), design mindset (ability to develop diverse tasks and workflows for positive outcomes), complex problem-solving skills (ability to proffer creative solutions), virtual collaboration (ability to work as part of a team to achieve desired outcomes), adaptive thinking skills (ability to think through problems), people, persuasion, and negotiating skills (ability to bargain with clients and colleagues), knowledge of advanced data analytics (knowledge of statistics and programming), and entrepreneurship skills (ability to develop marketable ideas) (Aulbur and Bigghe, 2016; Gray, 2016; Nafea and Toplu, 2020). This array of necessary skills is expected to compel universities to develop and revamp their curricula to develop the previously mentioned skills and competencies in the areas of engineering, management, and design and innovation. With regard to these discipline-specific domains, graduates will be expected to possess engineering knowledge such as big data analytics, advanced simulation, data communication, system automation, real-time inventory operations, artificial intelligence, robotics, cybersecurity, information technologies, augmented and virtual reality, and mechatronics (Xing and Marwala, 2017). They are also required to possess social skills (ability to connect to others deeply and directly), cross-cultural competence (ability to operate and adapt to different cultural settings), curiosity and imaginative skills, communication skills, and resource management skills (Motyl *et al.*, 2017).

According to Schwab (2017), there are four major categories of skills required for graduates to possess in the 4IR. They include workforce readiness, soft skills, technical skills, and entrepreneurship skills. Some of the skills peculiar to the workforce readiness attributes include personal presentation and management skills (time and self). The next set of skills are known as soft skills (essential human skills) and include attributes such as critical thinking skills, problem-solving skills, interpersonal relationships, communication skills, innovativeness, courage, adaptability, emotional intelligence (EI), and creativity. Technical skills include the knowledge required to thrive in one's workplace (targeted training) and can be developed among students via work-based learning. Finally, owing to the innovations peculiar to the 4IR, learners will be required to possess the ability to develop marketable ideas on their own without relying on the rigours of the recruitment processes. These four categories of skills provide the opportunity for lifelong learning required for the 4IR.

Are universities ready for 4IR?

The 4IR is not a concept for the future; it is upon us! As automation continues to advance rapidly, the pressures on universities to revamp and restructure their curricula (skills revolution) have increased. This can be achieved by introducing significant modifications to the science and technology curricula to enable students to develop competencies in the rapidly emerging areas of artificial intelligence, data science, robotics, advanced simulation, data communication, system automation, real-time inventory operations, cybersecurity, and information technologies. This implies that the curriculum should inculcate a 4IR element within the conventional primary sciences such as biology, chemistry, and physics, with greater emphasis on digital literacy to boost 4IR understanding among young learners. Specifically, within the biology discipline, new modules have been integrated to address emerging concepts such as innovative bioprocessing, genomic studies, synthetic biology, biotechnologies, and molecular design. For example, last year, a new biology module was introduced at the Stanford University in California that provides students with the opportunity to design real-time experiments to cure pathogens such as Lyme disease by means of what was termed 'using Robust B cell responses to predict rapid resolution of the disease'. This was achieved by the introduction of a new module in engineering biology that designs digital life forms to proffer practical solutions to problems arising in public health and other related sectors. Owing to this emerging biobased economy (bioeconomy), the Stanford curriculum has designed a major discipline called bioengineering which 'trains students at the interface of life sciences and engineering and merges expertise and resources in the departments of medicine, biology, and engineering (Gleason, 2018).

Apart from the biology scope, the chemistry field has also witnessed rapid changes as modules and programmes have become obtainable in green chemistry, 'a field which blends chemistry, biology, and environmental science to allow students to engage on real environmental problems such as synthetic fuels, bioplastics, and toxicology, and to train students in techniques to reduce pollution' (Xing and Marwala, 2017; Gleason, 2018). Similarly, a new physics curriculum has been designed to underline 4IR collaborative skills in which learners collaborate to conceptualize and invent sophisticated musical instruments and cryptographic gadgets.

From the foregoing discussion, universities around the world are required to accept the need for significant restructuring of curricula by integrating innovative science programmes and establishing interdisciplinary departments to accelerate the development of learners in the areas of artificial intelligence, biotechnology, data science, robotics, advanced simulation, nanotechnology materials, and several other technologies peculiar to the 4IR.

Several teaching methods can be considered for the training of learners in universities such as online courses, flipped modules, multidisciplinary approaches, and blended instructions to provide flexibility and modularity of modules (Gleason, 2018). Some universities in the United States have already begun the process of establishing effective blended learning environments for learners. One such recent example is the highly revered CS50 course, which is Harvard University's introduction to the intellectual enterprises of computer science and the art of programming. The course, which opened in January 2019, aims to:

> provide learners with a broad and robust understanding of computer science and programming; thinking algorithmically and solving programming problems efficiently; understanding many languages, including C, PHP, and JavaScript plus SQL, CSS, and HTML; engaging with a vibrant community of like-minded learners from all levels of experience and understanding concepts like abstraction, algorithms, data structures, encapsulation, resource management, security, software engineering, and web development.
>
> (Gleason, 2018)

Another example is the Introductory Electrical Engineering course which is currently being offered at the Massachusetts Institute of Technology (MIT). This course 'provides an integrated introduction to electrical engineering and computer science, taught using substantial laboratory experiments with mobile robots and also provides students with an understanding of modern software engineering, linear systems analysis, electronic circuits, and decision-making. In developing educational strategies in delivering 4IR knowledge to learners, Gleason (2018) offered a suggestion:

> any effective 4IR education strategy must also include in equal measure a deep consideration of the human condition, the ways in which new technologies and shifting economic power impact people of all socioeconomic levels, and the threats that exist within a world that is increasingly interconnected, in a way that fosters deep intercultural understanding and an abiding respect for freedom and human rights.

Outside the United States, several nations and their universities have introduced new modules and revamped exiting ones to fit into the 4IR. Some of these modules have been identified as key subjects that provide the foundations to learn certain 4IR components. For example, Statistical Methods of Data Analysis and Data Base Systems are courses that have been introduced into the curricula of the Turkish-German University, Turkey, as the basis of developing learners' idea of artificial intelligence and machine learning. To equip students ahead of 4IR challenges, the curricula consist of three specialization themes: hardware systems, information and technology (IT) security, and intelligent systems. For the hardware systems theme, modules such as robotics and automation technology are required. In the case of IT security, modules such as security management, cybersecurity, data privacy, and protocols are offered. Finally, the intelligent systems' theme includes modules such as artificial intelligence and machine learning.

Summary

This chapter extensively addressed the gaps observed in a bid to satisfy the purpose of this book. The critical examination and justification of these gaps form the basis upon which the conceptual framework for this book was achieved. In ensuring the preparedness of built environment graduates for the future of the South African construction industry and beyond, these two gaps and dimensions are pivotal in the proposed employability improvement framework. They are university–industry collaboration and the need for graduates to possess Industry 4.0 requirements. The next chapter will review the concept of learning and several ways to assess the learning outcomes of students. The chapter will also discuss teaching approaches and examine various teaching methods by which higher education can develop employability skills among students.

References

Abele, E., Metternich, J., Tisch, M., Chryssolouris, G., Sihn, W., El Maraghy, H., Hummel, V. and Ranz, F. 2015. Learning factories for research, education and training. *Procedia CIRP, 32*, pp. 1–6.

Adler, P. and Kwon, S. 2002. Social capital: Prospects for a new concept. *Academy of Management Review, 27*, pp. 17–40.

Ankrah, S. and Omar, A.T. 2015. Universities: Industry collaboration: A systematic review. *Scandinavian Journal of Management, 31*(3), pp. 387–408.

Archer, E. and Chetty, Y. 2013. Graduate employability: Conceptualisation and findings from the University of South Africa. *Progressio, 35*(1), pp. 136–167.

Armbrust, M., Fox, A., Griffith, R., Joseph, A.D., Katz, R., Konwinski, A., Lee, G., Patterson, D., Rabkin, A., Stoica, I. and Zaharia, M. 2010. A view of cloud computing. *Communications of the ACM, 53*(4), pp. 50–58.

Artess, J., Forbes, P. and Ripmeester, N. 2011. *Supporting graduate employability: HEI practice in other countries.* BIS Research Paper Number 40. London: BIS.

Aulbur, W. and Bigghe, R. 2016. *Skill development for industry 4.0: BRICS skill development working group.* Munich: Roland Berger GMBH.

Bahrin, M.A.K., Othman, M.F., Azli, N.N. and Talib, M.F. 2016. Industry 4.0: A review on industrial automation and robotic. *Jurnal Teknologi, 78*(6–13), pp. 137–143.

Barnes, T., Pashby, I. and Gibbons, A. 2002. Effective university-industry interaction: A multi-case evaluation of collaborative R&D projects. *European Management Journal, 20*, pp. 272–285.

Barringer, B.R. and Harrison, J.S. 2000. Walking a tightrope: Creating value through inter-organisational relationships. *Journal of Management, 26*(3), pp. 367–403.

Beckett, D. and Mulcahy, D. 2006. Constructing professionals' employ-abilities: Conditions for accomplishment. In P. Hager and S. Holland (Eds.), *Graduate attributes, learning and employability* (pp. 243–265). Dordrecht, the Netherlands: Springer.

Bekkers, R. and Bodas Freitas, I. 2008. Analysing knowledge transfer channels between universities and industry: To what degree do sectors also matter? *Research Policy, 37*, pp. 1837–1853.

Bettis, R. and Hitt, M. 1995. The new competitive landscape. *Strategic Management Journal, 16*, p. 719.

Bezuidenhout, M. 2011. *The development and evaluation of a measure of graduate employability in the context of the new world of work* (Doctoral dissertation, University of Pretoria, Pretoria).

Blau, P.M. 1977. *Inequality and heterogeneity: A primitive theory of social structure* (Vol. 7). New York: Free Press.

Boddy, D., Macbeth, D. and Wagner, B. 2000. Implementing collaboration between organizations: An empirical study of supply chain partnering. *Journal of Management Studies*,
37, pp. 1003–1018.

Bouri, A., Breij, M., Diop, M., Kempner, R., Klinger, B. and Stevenson, K. 2011. *Report on
support to SMEs in developing countries through financial intermediaries*. New York:
Dalberg.

Bracht, U. and Masurat, T. 2005. The digital factory between vision and reality. *Computers
in Industry*, 56(4), pp. 325–333.

Brass, D.J., Galaskiewicz, J., Greve, H.R. and Tsai, W. 2004. Taking stock of networks and
organizations: A multilevel perspective. *Academy of Management Journal*, 47(6),
pp. 795–817.

Brettel, M., Friederichsen, N., Keller, M. and Rosenberg, M. 2014. How virtualization,
decentralization and network building change the manufacturing landscape: An industry 4.0 perspective. *International Journal of Mechanical, Industrial Science and Engineering*,
8(1), pp. 37–44.

Bridgstock, R. 2009. The graduate attributes we have overlooked: Enhancing graduate
employability through career management skills. *Higher Education Research & Development*, 28(1), pp. 31–44. doi: 10.1080/07294360802444347.

Brockner, J. 1988. The effects of work layoffs on survivors: Research, theory, and practice.
In B.M. Staw and L.L. Cummings (Eds.), *Research in organizational behavior* (Vol. 10,
pp. 213–255). Greenwich, CT: JAI Press.

Brynjolfsson, E. and McAfee, A. 2014. *The second machine age: Work, progress, and prosperity in a time of brilliant technologies*. New York, NY: WW Norton & Company.

Burkhardt, M.E. and Brass, D.J. 1990. Changing patterns or patterns of change: The effect
of a change in technology on social network structure and power. *Administrative Science
Quarterly*, 35, pp. 104–127.

Cachay, J., Wennemer, J., Abele, E. and Tenberg, R. 2012. Study on action-oriented learning with a learning factory approach. *Procedia-Social and Behavioural Sciences*, 55,
pp. 1144–1153.

Casciaro, T. 1998. Seeing things clearly: Social structure, personality, and accuracy in
social network perception. *Social Networks*, 20, pp. 331–351.

Child, J., Faulkner, D. and Tallman, S. 2005. *Cooperative strategy: Managing alliances, networks, and joint ventures*. Oxford: Oxford University Press.

Coetzee, M. 2012. A framework for developing student graduateness and employability in
the economic and management sciences at the University of South Africa. *Developing
student graduateness and employability: Issues, provocations, theory and practical guidelines*
(pp. 119–152). Randburg: Copyright © Knowres Publishing.

Cohen, W.M., Florida, R., Randazzese, L. and Walsh, J. 1998. Industry and the academy:
Uneasy partners in the cause of technological advance. In R. Noll (Ed.), *The future of
the research university*. Washington, DC: Brookings Institution Press.

Cohen, W.M. and Levinthal, D.A. 1990. Absorptive capacity: A new perspective on learning and innovation. *Administrative Science Quarterly*, 35, pp. 128–152.

Dacin, M.T., Oliver, C. and Roy, J.-P. 2007. The legitimacy of strategic alliances: An institutional perspective. *Strategic Management Journal*, 28, pp. 169–187.

Dacre Pool, L. and Sewell, P. 2007. The key to employability: Developing a practical model
of graduate employability. *Education Training*, 49(4), pp. 277–289. https://doi.
org/10.1108/00400910710754435.

Danowski, J.A. and Edison-Swift, P. 1985. Crisis effects on interorganisational computer-
based communication. *Communication Research*, 12(2), pp. 251–270.

Dekker, H.C. 2004. Control of inter-organizational relationships: Evidence on appropriation concerns and coordination requirements. *Accounting, Organizations and Society, 29*, pp. 27–49.

DiMaggio, P. and Powell, W. 1983. The iron cage revisited: Institutional isomorphism and collective rationality in organizational fields. *American Sociological Review, 48*, pp. 147–160.

Du Plessis, C.J. 2017. *A framework for implementing industry 4.0 in learning factories* (Doctoral dissertation, Stellenbosch University, Stellenbosch).

Dzisah, J. and Etzkowitz, H. 2008. Triple helix circulation: The heart of innovation and development. *International Journal of Technology Management and Sustainable Development, 7*(2), pp. 101–115.

Eames, C. and Cates, C. 2011. Theories of learning in cooperative and work-integrated education. In R.K. Coll and K.E. Zegwaard (Eds.), *International handbook for cooperative and work-integrated education* (2nd ed., pp. 41–52). Lowell, MA: World Association for Cooperative Education Inc.

Elder, S. 2014. *Labour market transitions of young women and men in Asia and the Pacific.* Geneva, Switzerland: International Labour Office.

Etzkowitz, H. 2008. *The triple helix: University-industry-government innovation in action.* New York: Routledge.

Etzkowitz, H. and Leydesdorff, L.A. 2000. The dynamics of innovation: From national systems and mode 2 to a triple helix of university-industry-government relations. *Research Policy, 29*(2), pp. 109–123.

Evans, D. 2011. The internet of things: How the next evolution of the internet is changing everything. *CISCO White Paper, 1*(2011), pp. 1–11.

Ferns, S., Campbell, M. and Zegwaad, K. 2014. Work integrated learning. In S. Ferns (Ed.), *HERDSA guide: Work integrated learning in the curriculum* (pp. 1–6). Milperra, NSW: Higher Education Research and Development Society of Australasia.

Ferns, S. and Moore, K. 2012. Assessing student outcomes in fieldwork placements: An overview of current practice. *Asia-Pacific Journal of Cooperative Education, 13*(4), pp. 207–224.

Fongwa, S. 2018. Towards an expanded discourse on graduate outcomes in South Africa. *Education as Change, 22*(3), pp. 1–23.

Freeman, C. and Soete, L. 1997. *The economics of industrial innovation.* London and New York: Routledge Taylor and Francis.

Freeman, R.E. 1994. *Ethical theory and business.* Englewood Cliffs, NJ: Prentice-Hall.

Fugate, M. and Ashforth, B. 2004. Employability: The construct, its dimensions, and applications. *Proceedings of the Annual Meeting of the Academy of Management, OB: J1–J6*, Seattle.

Gani, K. 2017. *Employability attributes organisational commitment and retention factors in the 21st century world of work* (Doctoral dissertation, University of South Africa, Pretoria).

Gasser, M. 1988. *Building a secure computer system* (p. 85). New York: Van Nostrand Reinhold Company.

Geel, M. 2014. *An investigation into the employability skills of undergraduate Business Management students.* Potchefstroom: North-West University.

Geisler, E. 1995. Industry-university technology cooperation: A theory of inter-organizational relationships. *Technology Analysis & Strategic Management, 7*, pp. 217–229.

Geissbauer, R., Vedso, J. and Schrauf, S. 2016. *Industry 4.0: Building the digital enterprise.* Retrieved from: www.pwc.com/gx/en/industries/industries-4.0/landing-page/industry-4.0-building-your-digital-enterprise-april-2016.pdf [Accessed 25 December 2016].

George, G., Zahra, S.A. and Wood, D.R. 2002. The effects of business-university alliances on innovative output and financial performance: A study of publicly traded biotechnology companies. *Journal of Business Venturing, 17*, pp. 577–609.

Gerwin, D. and Tarondeau, J.C. 1982. Case studies of computer integrated manufacturing systems: A view of uncertainty and innovation processes. *Journal of Operations Management, 2*(2), pp. 87–99.

Gleason, N.W. (Ed.). 2018. *Higher education in the era of the fourth industrial revolution.* Singapore: Springer Nature.

Goldhar, J.D. and Jelinek, M. 1983. Plan for economies of scope. *Harvard Business Review, 61*(6), pp. 141–148.

Gray, A. 2016, January. The 10 skills you need to thrive in the fourth industrial revolution. In *World Economic Forum, 19.*

Gregor, M., Medvecky, S., Matuszek, J. and Stefánik, A. 2009. Digital factory. *Journal of Automation Mobile Robotics and Intelligent Systems, 3*, pp. 123–132.

Griesel, H. and Parker, B. 2009. Graduate attributes. *Higher Education South Africa and the South African Qualifications Authority.* Pretoria.

Hagan, D. 2004. Employer satisfaction with ICT graduates. *Proceedings of the Sixth Australasian Conference on Computing Education*, Australian Computer Society, Inc, Sydney, Vol. 30, pp. 119–123.

Hakansson, H. and Snehota, I. (Eds.). 1995. *Developing relationships in business networks.* London: Routledge.

Hamel, G., Doz, Y.L. and Prahalad, C.K. 1989. Collaborate with your competitors and win. *Harvard Business Review, 67*(1), pp. 133–139.

Harbison, J.R. and Pekar, P., Jr. 1998. *Smart alliances: A practical guide to repeatable success.* Jossey-Bass.

Harman, G. and Sherwell, V. 2002. Risks in university-industry research links and the implications for university management. *Journal of Higher Education Policy and Management, 24*, pp. 37–51.

Haupt, T. 2012. Student attitudes towards cooperative construction education experiences. *Construction Economics and Building, 3*(1), pp. 31–42.

Hecklau, F., Galeitzke, M., Flachs, S. and Kohl, H. 2016. Holistic approach for human resource management in industry 4.0. *Procedia CIRP, 54*, pp. 1–6.

Helyer, R., Lee, D. and Evans, A.S. 2011. Hybrid HE: Knowledge, skills and innovation. *Work Based Learning e-Journal, 1*(2), pp. 18–34.

Henning, K., Wolfgang, W. and Johannes, H. 2013. *Recommendations for implementing the strategic initiative INDUSTRIE 4.0.* Final report of the Industrie (Vol. 4, p. 82).

Hermann, M., Pentek, T. and Otto, B. 2016, January. Design principles for Industrie 4.0 scenarios. *Proceedings of the System Sciences (HICSS), 2016 49th Hawaii International Conference*, IEEE, Annual Hawaii International Conference on System Sciences (HICSS), Hawaii, pp. 3928–3937.

Hilbert, M. 2013. *Big data for development: From information- to knowledge societies.* Retrieved from SSRN: https://ssrn.com/abstract=2205145 or http://dx.doi.org/10.2139/ssrn.2205145.

Hobsbawm, E. 2016. *Da revolução industrialinglesa ao imperialismo* (6th ed.). Rio de Janeiro: Forense Universitária.

Hoffmann, W. and Schlosser, R. 2001. Success factors of strategic alliances in small and medium-sized enterprises, an empirical survey. *Long Range Planning, 34*, pp. 357–381.

Hong, W. and Su, Y.-S. 2013. The effect of institutional proximity in non-local university-industry collaborations: An analysis based on Chinese patent data. *Research Policy, 42*, pp. 454–464.

Huber, M.T. and Hutchings, P. 2004. *Integrative learning: Mapping the terrain: The academy in transition*. Washington, DC: Association of American Colleges and Universities and the Carnegie Foundation for the Advancement of Teaching.

Ibarra, H. 1993. Personal networks of women and minor cities in management: A conceptual framework. *Academy of Management Review*, *18*, pp. 56–87.

Jonck, P. 2014. A human capital evaluation of graduates from the Faculty of Management Sciences employability skills in South Africa. *Academic Journal of Interdisciplinary Studies*, *3*(6), pp. 265–274. https://doi.org/10.5901/ajis.2014.v3n6p265.

Kagermann, H., Wahlster, W. and Helbig, J. 2013. *Recommendations for implementing the strategic initiative industrie 4.0*. Final report of the Industrie 4.0 working group.

Kamble, S.S., Gunasekaran, A. and Gawankar, S.A. 2018. Sustainable industry 4.0 framework: A systematic literature review identifying the current trends and future perspectives. *Process Safety and Environmental Protection*, *117*, pp. 408–425.

Kanter, R.M. 1977. *Men and women of the corporation*. New York: Basic Books.

Kharazmi, O.A. 2011. *Modelling the role of university-industry collaboration in the Iranian national system of innovation: Generating transition policy scenarios*. Stirling: University of Stirling.

Klein, K.J., Lim, B.C., Saltz, J.L. and Mayer, D.M. 2004. How do they get there? An examination of the antecedents of centrality in team networks. *Academy of Management Journal*, *47*(6), pp. 952–963.

Kravchenko, A. and Kyzymenko, I. 2019. The fourth industrial revolution: New paradigm of society development or posthumanist manifesto. *Philosophy & Cosmology*, *22*, pp. 120–128.

Kuhn, W. 2006, December. Digital factory-simulation enhancing the product and production engineering process. *Proceedings of the Simulation Conference*, *2006*, WSC 06, IEEE, Piscataway, NJ, pp. 1899–1906.

Kundaeli, F. 2016. *Individual factors affecting the employability of information systems graduates in Cape Town, South Africa: Employed graduates and employer perspectives* (Doctoral dissertation, University of Cape Town, Cape Town).

Larsson, R., Bengtsson, L., Henriksson, K. and Sparks, J. 1998. The interorganisational learning dilemma: Collective knowledge development in strategic alliances. *Organization Science*, *9*, pp. 285–305.

Law, W. and Watts, A.G. 1977. *Schools, careers and community*. London: Church Information Office.

Lee, E.A. 2010, June. CPS foundations. *Proceedings of the 47th Design Automation Conference*, ACM Press, New York, NY, pp. 737–742

Lee, J., Bagheri, B. and Kao, H.A. 2015. A cyber-physical systems architecture for industry 4.0-based manufacturing systems. *Manufacturing Letters*, *3*, pp. 18–23.

Lee, Y. 2000. The sustainability of university-industry research collaboration: An empirical assessment. *Journal of Technology Transfer*, *25*, pp. 111–133.

Lei, D., Hitt, M.A. and Goldhar, J.D. 1996. Advanced manufacturing technology: Organizational design and strategic flexibility. *Organization Studies*, *17*(3), pp. 501–523.

Levinthal, D. and March, J.G. 1993. The myopia of learning. *Strategic Management Journal*. Winter Special Issue, *14*, pp. 95–112.

Liao, Y., Deschamps, F., Loures, E.D.F.R. and Ramos, L.F.P. 2017. Past, present and future of industry 4.0: A systematic literature review and research agenda proposal. *International Journal of Production Research*, *55*(12), pp. 3609–3629.

Liao, Y., Loures, E.R., Deschamps, F., Brezinski, G. and Venâncio, A. 2018. The impact of the fourth industrial revolution: A cross-country/region comparison. *Production*, *28*.

Lin, C.C., Deng, D.J., Chen, Z.Y. and Chen, K.C. 2016. Key design of driving industry 4.0: Joint energy-efficient deployment and scheduling in group-based industrial wireless sensor networks, *Institute of Electrical and Electronics Engineers Communication Magazine*, 54(10).

Liu, Y. and Xu, X. 2017. Industry 4.0 and cloud manufacturing: A comparative analysis. *Journal of Manufacturing Science and Engineering*, 139(3), p. 034701.

Logar, C.M., Ponzurick, T.G., Spears, J.R. and France, K.R. 2001. Commercializing intellectual property: A university-industry alliance for new product development. *Journal of Product and Brand Management*, 10, pp. 206–217.

Lowden, K., Hall, S., Ellio, D.D. and Lewin, J. 2011. *Employers' perceptions of the employability skills of new graduates*. London: Edge Foundation.

Lucke, D., Constantinescu, C. and Westkämper, E. 2008. Smart factory: A step towards the next generation of manufacturing. In *Manufacturing systems and technologies for the new frontier* (pp. 115–118). London: Springer.

Luthra, S. and Mangla, S.K. 2018. Evaluating challenges to industry 4.0 initiatives for supply chain sustainability in emerging economies. *Process Safety and Environmental Protection*, 117, pp. 168–179.

Madakam, S., Ramaswamy, R. and Tripathi, S. 2015. Internet of Things (IoT): A literature review. *Journal of Computer and Communications*, 3(5), p. 164.

Manyika, J., Chui, M., Brown, B., Bughin, J., Dobbs, R., Roxburgh, C. and Byers, A.H. 2011. *Big data: The next frontier for innovation, competition, and productivity*. Washington, DC: McKinsey Global Institute.

McPherson, M. 1983. An ecology of affiliation. *American Sociological Review*, pp. 519–532.

McQuaid, R.W. and Lindsay, C. 2005. The concept of employability. *Urban Studies*, 42(2), pp. 197–219.

Mehra, A., Kilduff, M. and Brass, D.J. 2001. The social networks of high and low self-monitors: Implications for workplace performance. *Administrative Science Quarterly*, 46, pp. 121–146.

Mell, P. and Grance, T. 2011. *The NIST definition of cloud computing*. NIST Special Publication 800-145. Gaithersburg, MD: NIST.

Mitchell, W. and Singh, K. 1996. Survival of businesses using collaborative relationships to commercialize complex goods. *Strategic Management Journal*, 17, pp. 169–195.

Monge, P.R. and Eisenberg, E.M. 1987. Emergent communication networks. In F.M. Jablin, L.L. Putnam, K.H. Roberts and L.W. Porter (Eds.), *Handbook of organizational communication: An interdisciplinary perspective* (pp. 304–342). Newbury Park, CA: Sage Publications.

Mora-Valentin, E.M. 2000. University-industry cooperation: A framework of benefits and obstacles. *Industry and Higher Education*, 14, pp. 165–172.

Moreno, A., Velez, G., Ardanza, A., Barandiaran, I., De Infante, Á.R. and Chopitea, R. 2017. Virtualisation process of a sheet metal punching machine within the industry 4.0 vision. *International Journal on Interactive Design and Manufacturing (IJIDeM)*, 11(2), pp. 365–373.

Motyl, B., Baronio, G., Uberti, S., Speranza, D. and Filippi, S. 2017. How will change the future engineers' skills in the industry 4.0 framework? A questionnaire survey. *Procedia Manufacturing*, 11, pp. 1501–1509.

Mtebula, C.T. 2015. *Employers' and graduates perception survey on employability and graduateness: Products of the School of Construction Economics and Management at the University of the Witwatersrand* (Doctoral dissertation, University of the Witwatersrand, Johannesburg).

Nafea, R.M.E.D. and Toplu, E.K. 2020. Future of education in industry 4.0: Educational digitization: A Canadian case study. In *Business management and communication perspectives in industry 4.0* (pp. 267–287). Hershey, PA: IGI Global.

Naicker, D. 2017. *Tourism graduate employability: Stakeholder perceptions of workplace learning for graduate employment* (Doctoral dissertation, Durban University of Technology, Durban).

Ndzube, F. 2013. *The relationship between career anchors and employability* (Doctoral dissertation, University of South Africa, Pretoria).

Newberg, J.A. and Dunn, R.L. 2002. Keeping secrets in the campus lab: Law, values and rules of engagement for industry-university R&D partnerships. *American Business Law Journal*, 39, pp. 187–241.

Oesterreich, T.D. and Teuteberg, F. 2016. Understanding the implications of digitisation and automation in the context of industry 4.0: A triangulation approach and elements of a research agenda for the construction industry. *Computers in Industry*, 83, pp. 121–139.

Oliver, C. 1990. Determinants of interorganisational relationships: Integration and future directions. *Academy of Management Review*, 15, pp. 241–265.

Pereira, A.C. and Romero, F. 2017. A review of the meanings and the implications of the industry 4.0 concept. *Procedia Manufacturing*, 13, pp. 1206–1214.

Perkmann, M., King, Z. and Pavelin, S. 2011. Engaging excellence, effects of faculty quality on university engagement with industry. *Research Policy*, 40, pp. 539–552.

Perkmann, M., Tartari, V., McKelvey, M., Autio, E., Broström, A., D'Este, P., Fini, R., Geuna, A., Grimaldi, R., Hughes, A. and Krabel, S. 2013. Academic engagement and commercialisation: A review of the literature on university-industry relations. *Research Policy*, 42(2), pp. 423–442.

Pfeffer, J. and Salancik, G. 1978. *The external control of organizations: A resource dependence perspective*. New York, NY: Harper and Row.

Plass, C. 2015. *Facts for decision makers: Seize the opportunity of industry 4.0.*

Potgieter, I.L. 2013. *Development of a career meta-competency model for sustained employability* (Doctoral dissertation, University of South Africa, Pretoria).

Powell, W.W., Koput, K.W. and Smith-Doerr, L. 1996. Interorganisational collaboration and the locus of innovation: Networks of learning in biotechnology. *Administrative Science Quarterly*, 41, pp. 116–145.

Qin, J., Liu, Y. and Grosvenor, R. 2016. A categorical framework of manufacturing for industry 4.0 and beyond. *Procedia CIRP*, 52, pp. 173–178.

Ramadan, M., Al-Maimani, H. and Noche, B. 2017. RFID-enabled smart real-time manufacturing cost tracking system. *The International Journal of Advanced Manufacturing Technology*, 89(1–4), pp. 969–985.

Ramakrishnan, K. and Yasin, N.M. 2011. Higher learning institution-industry collaboration: A necessity to improve teaching and learning process. *Proceedings of the Computer Science & Education (ICCSE) 2011 6th International Conference on IEEE*, Singapore, p. 1445.

Rüssmann, M., Lorenz, M., Gerbert, P., Waldner, M., Justus, J., Engel, P. and Harnisch, M. 2015. *Industry 4.0: The future of productivity and growth in manufacturing industries* (p. 9). Boston, MA: Boston Consulting Group.

Rust, C. and Froud, L. 2011. Personal literacy: The vital, yet often overlooked, graduate attribute. *Journal of Teaching and Learning for Graduate Employability*, 2(1), pp. 28–40.

Saad, M. and Zawdie, G. 2005. From technology transfer to the emergence of a triple helix culture: The experience of Algeria in innovation and technological capability development. *Technology Analysis and Strategic Management, 17*(1), pp. 89–103.

Sackey, S.M., Bester, A. and Adams, D. 2017. Industry 4.0 learning factory didactic design parameters for industrial engineering education in South Africa. *South African Journal of Industrial Engineering, 28*(1), pp. 114–124.

Santoro, M.D. and Chakrabarti, A.K. 2001. Corporate strategic objectives for establishing relationships with university research centres. *IEEE Transactions on Engineering Management, 48*, pp. 157–163.

Santos, F.M. and Eisenhardt, K.M. 2005. Organizational boundaries and theories of organization. *Organization Science, 16*, pp. 491–508.

Schartinger, D., Rammer, C. and Fröhlich, J. 2006. Knowledge interactions between universities and industry in Austria: Sectoral patterns and determinants. In *Innovation, networks, and knowledge spillovers* (pp. 135–166). Berlin: Springer.

Schwab, K. 2017. *The fourth industrial revolution.* Currency.

Scott, J. 1987. *Organizations.* Englewoods Cliffs, NJ: Simon and Schuster.

Shah, P. 2000. Network destruction: The structural implications of downsizing. *Academy of Management Journal, 43*, pp. 101–112.

Sherwood, A.L., Butts, S.B. and Kacar, S.L. 2004, October. Partnering for knowledge: A learning framework for university-industry collaboration. *Midwest Academy of Management, 2004 Annual Meeting*, New Orleans, LA, pp. 1–17.

Shivoro, R.S., Shalyefu, R.K. and Kadhila, N. 2018. Perspectives on graduate employability attributes for management sciences graduates. *South African Journal of Higher Education, 32*(1), pp. 216–232.

Siegel, D.S., Waldman, D.A., Atwater, L.E. and Link, A.N. 2003. Commercial knowledge transfers from universities to firms: Improving the effectiveness of university-industry collaboration. *The Journal of High Technology Management Research, 14*(1), pp. 111–133.

Siegel, D.S., Waldman, D.A., Atwater, L.E. and Link, A.N. 2004. Toward a model of the effective transfer of scientific knowledge from academicians to practitioners: Qualitative evidence from the commercialization of university technologies. *Journal of Engineering and Technology Management, 21*(1–2), pp. 115–142.

Siemieniuch, C.E. and Sinclair, M.A. 2015. Global drivers, sustainable manufacturing and systems ergonomics. *Applied Ergonomics, 51*, pp. 104–119.

Simonin, B.L. 1997. The importance of collaborative know-how: An empirical test of the learning organization. *Academy of Management Journal, 40*, pp. 1150–1174.

Sternberg, R.J., Forsythe, G.B., Hedlund, J., Wagner, R.K., Horvath, J.A., Williams, W.M., Snook, S.A. and Grigorenko, E. 2000. *Practical intelligence in everyday life.* Cambridge: Cambridge University Press.

Sundmaeker, H., Guillemin, P., Friess, P. and Woelfflé, S. 2010. Vision and challenges for realising the Internet of Things. *Cluster of European Research Projects on the Internet of Things, European Commission, 3*(3), pp. 34–36.

Symington, N. 2012. *Investigating graduate employability and psychological career resources* (Doctoral dissertation, University of Pretoria, Pretoria).

Tadelis, S. and Williamson, O. 2012. Transaction cost economics. In R. Gibbons and J. Roberts (Eds.), *the handbook of organizational economics.* Princeton, NJ: Princeton University Press.

Terzidis, O., Oberle, D., Friesen, A., Janiesch, C. and Barros, A. 2012. The internet of services and USDL. In *Handbook of service description* (pp. 1–16). Boston, MA: Springer.

Tran, T.T. 2016. Enhancing graduate employability and the need for university-enterprise collaboration. *Journal of Teaching and Learning for Graduate Employability, 7*(1), pp. 58–71.

Van Dam, K. 2004. Antecedents and consequences of employability orientation. *European Journal of Work and Organizational Psychology, 13*(1), pp. 29–51.

Wagner, U., Al Geddawy, T., El Maraghy, H. and MŸller, E. 2012. The state-of-the-art and prospects of learning factories. *Procedia CIRP, 3*, pp. 109–114.

Weisz, M. and Smith, S. 2010. Critical changes for successful cooperative education. In A. Brew and C. Asmar (Eds.), *Higher education in a changing world: Research and development in higher education* (Vol. 28, pp. 605–615). Sydney, NSW: Higher Education Research and Development Society of Australasia (HERDSA).

Winer, M. and Ray, K. 1994. *Collaboration handbook: Creating, sustaining, and enjoying the journey*. St. Paul, MN: Amherst H. Wilder Foundation.

Wright, M., Clarysseb, B., Lockett, A. and Knockaertd, M. 2008. Midrange universities' linkages with industry: Knowledge types and the role of intermediaries. *Research Policy, 37*, pp. 1205–1223.

Xing, B. and Marwala, T. 2017. Implications of the fourth industrial age for higher education. *The_Thinker__Issue_73__Third_Quarter_2017*.

Yorke, M. and Knight, P. 2006. *Embedding employability into the curriculum*. York: Higher Education Academy.

Zegwaard, K.E. and Coll, R.K. 2011. Using cooperative education and work-integrated education to provide career clarification. *Science Education International, 22*(4), pp. 282–291.

Zhang, Z., Liu, S. and Tang, M. 2014. Industry 4.0: Challenges and opportunities for Chinese manufacturing industry. *Technical Gazette, 21*(6), pp. III–IV.

Zhou, W., Piramuthu, S., Chu, F. and Chu, C. 2017. RFID-enabled flexible warehousing. *Decision Support Systems, 98*, pp. 99–112.

Zukin, S. and Dimaggio, P. 1990. *Structures of capital: The social organization of the economy*. New York: Cambridge University Press.

4 Learning theories and skills development

Theories of learning

Overview of learning

The art of human learning is a continuous activity until one dies. This has prompted researchers from various academic backgrounds and traditions to continue testing their theories and assumptions in various settings as the quest to improve learning and teaching continues. There is a general belief that learning is important; however, a general definition has been difficult to arrive at owing to different views regarding the methods and consequences of learning (Shuell, 1986). The concept of human learning reflects the process of effective interaction with the immediate environment (both physical and social). However, the definition by Schunk (2012) has been able to reflect the thinking of researchers, theorists, educators, and industry stakeholders. According to Schunk (2012), learning is 'an enduring change in behaviour, or in the capacity to behave in a given fashion, which results from practice or other forms of experience' (Schunk, 2012). This definition provides three critical indicators of learning: change, evolution, and experience. The first indicator of change stipulates that individuals can effectively learn when they attain the ability of doing something differently.

The second indicator of evolution stipulates that individuals can be conversant with taught ideas, but may forget as time goes on. The third indicator of experience stipulates that individuals can gain understanding through continuous practice and observation from other people (learners and teachers) (Schunk, 2012). Another critical definition as presented by Kimble (1964) describes learning as 'a relatively permanent change in behavioural potentiality that occurs as a result of reinforced practice'. These definitions highlight several key terms in its definition that make them pivotal to this study. The first term, namely 'relatively permanent', suggests that the behavioural changes learning brings about are relatively permanent (long-term) and not temporary. The 'behavioural change' aspect suggests that effective learning must result in a significant and visible change in behaviour which was not originally there before learning occurred. The 'potential' aspect of the definition suggests that the learning process may not instantaneously result in behavioural change, but will occur over a period. The 'reinforced' aspect suggests the thoroughness of learning activities and processes, while the 'practice' aspect suggests that behavioural change is a function of rigorous training and practice (Kimble, 1964). From Kimble's definition, a revised definition posited by Hergenhahn (1988) describes learning as 'a relatively permanent change in behaviour or in behavioural potentiality that results from experience and cannot be attributed to temporary body states such as those induced by illness, fatigue, or drugs' (Hergenhahn, 1988).

Furthermore, Stirling *et al.* (2016) presented four principal modes of learning: concrete experience (CE), reflective observation (RO), abstract conceptualization (AC), and active experimentation (AE). Concrete experience underscores what an individual has experienced and highlights the feelings of individuals in present reality. Individuals who are inclined to this learning mode are sensitive, highly interactive, and can adapt to different types of environment. The next mode of learning, namely reflective observation, describes the observations of the experience. Individuals who engage in reflection tend to examine how and what prompted the occurrence of given events. Individuals inclined to this mode of learning are versatile and exercise considerate judgement at any given time. Abstract conceptualization underlines the need to apply logic when describing the concept of experience, hence it primarily considers a scientific approach. Individuals inclined to this mode of learning are meticulous thinkers and systematic critiques of concepts. Lastly, the active experimentation underlines the importance of experimentation (creation of practical approaches) to influence an experience (solve arising issues). Individuals inclined to this mode of learning are perpetual risk-takers when seeking solutions to problems (Stirling *et al.*, 2016).

Besides learning modes, when acquiring new or building on existing knowledge, four basic learning styles are often considered. These four learning styles are converging, diverging, assimilating, and accommodating. Each of the learning styles is inclined towards at least two learning modes. The converging style of learning is associated with both abstract conceptualization and active experimentation. The main forms of skills and competencies most associated with the converging learning style are the abilities to solve problems and reason productively. The diverging style of learning, on the other hand, is associated with both reflective observation and concrete experience. The main form of skills and competencies most associated with the diverging learning style is the ability to be creative when observing various perspectives. In addition, the assimilating style of learning is associated with both reflective observation and abstract conceptualization. The main form of skills and competencies most associated with the assimilating learning style is the ability to interpret abstract ideas. Lastly, the accommodating style of learning is associated with both the active experimentation and concrete experience. The main forms of skills and competencies most associated with the accommodating learning style are the ability to adapt to new thinking, take risks, and actively engage in productive activities (Kolb, 1984). In evaluating the level of learning among students, researchers and stakeholders responsible for student training can assess their assimilation through learning outcomes. This is because learning is inferential and cannot be measured directly but rather through its outcomes and products. This leads to the issue of assessment. According to Popham (2008), assessment refers to a formal attempt adopted to ascertain the level of educational learning attained by students. There are several ways to assess the learning outcomes for students. They include direct observations, oral and written responses, assessment by others, and self-reports (Schunk, 2012).

Direct observations

Direct observation is an assessment style regularly employed by educators (teachers) which involves the physical assessment of how learning has occurred among students. For instance, a physics teacher who wants students to understand laboratory procedures ensures regular check-ups during classes to ensure they handle the apparatuses and laboratory equipment properly. Similarly, a sports instructor or physical and health education teacher constantly observes students who are learning a new sporting code or technique to assess how they are learning the skills of that sport in particular. This assessment style is a valid learning evaluator if they are as candid as possible and involve little or no interference from third parties. Notwithstanding the effectiveness of this assessment style, direct observations have their share of challenges. They focus more on the visible outcomes of learning and focus less on the possible cognitive and affective processes that trigger certain behaviours. For example, a physics teacher who knows that students have understood the laboratory functions and procedures may not be aware of their thoughts while performing an experiment. While learning differs from performance, several factors can play a significant role in affecting the performance of students. Factors such as state of mind, health, personal issues, and lack of motivation can also affect the performance of students. Hence, direct observation may not necessarily tell the whole story (Popham, 2008; Schunk, 2012).

Oral responses

Oral responses are an assessment style regularly employed by educators to determine the level of understanding among students by randomly asking them questions during lectures. In this style, the manner of questions asked of students is also an indicator of their level of learning and understanding. While this style may reflect the understanding of students, it is fair to say that several factors may prevent oral responses from being the perfect student-assessment tool. Factors such as language barriers and difficulties, stage fright (anxiety), and unfamiliar terminology may affect the translation of a student's idea into spoken words or responses (Popham, 2008; Schunk, 2012).

Written responses

Written responses are an assessment style regularly employed by educators to determine the level of understanding among students by designing quizzes, take-home assignments, report writing, and term papers. In this style, the educators determine the level of learning and understanding by students based on their responses and mastery of instructions in solving these written examination materials. Like the oral response style, different factors can prevent this type of assessment from being a perfect tool. Several factors such as illness and cheating can affect the written response of students and may not fully reflect what they have learned (Popham, 2008; Schunk, 2012).

Assessment by others

Assessment by others is an assessment style regularly employed by several individuals (peers, parents, researchers) in ascertaining the level of understanding and quality of learning attained by students. It can take the form of inquisitive questions that can help in understanding and identifying students who require extra motivation and lessons. In this style, the observation and assessment by others may be more unbiased about learners than the learners are about themselves (Popham, 2008; Schunk, 2012).

Self-reports

Self-reports are an assessment style regularly conducted by individuals themselves. It can take various forms, including dialogues, interviews, questionnaires, stimulated recollections, and thinking aloud. Dialogues are a random or intentional conversation between individuals while carrying out a learning activity. These conversations can be documented and evaluated in detail to determine factors and possible behavioural patterns that can affect learning. A questionnaire involves obtaining various items of information (thoughts) from respondents by means of designing a set of questions (open-ended or closed-ended). The responses can be used to measure their level of competence, types of activities, frequency of actions, and other information that is germane to understanding their level of learning. In addition, interviews are forms of questionnaire that are conducted individually or in groups, allowing respondents to discuss answers orally. In stimulated recollections, on the other hand, individuals carry out a task and later attempt to remember their thought processes that occurred at various levels of points during the activity. The assessment technique must be conducted soon after the learning task so that individuals do not forget their thought processes. In the thinking aloud process, individuals voice out their feelings, thoughts, and queries while working on a given learning task. These verbalizations may be written down, recorded, or documented by educators and observers and then analysed to determine the level of understanding among students. While some individuals struggle to speak out while working, observers will be required to motivate students to explain their thought processes audibly (Popham, 2008; Schunk, 2012). According to Ertmer and Newby (2013), the concept of learning is classified into two aspects: the behavioural and cognitive theory and the constructivist theory. Both of these perspectives of learning share common boundaries but are idiosyncratic enough to be considered as stand-alone approaches to underpin the concept of learning. To obtain a holistic view of the learning concept, the neurophysiological theories and humanist theories will also be discussed.

Behaviourism and cognitive learning theory

The concept of learning has always been referred to as the accumulation of several connected and sequential bits of valuable information that needs to be understood before being disseminated to learners (Earl and Katz, 2006). The behaviourism theory provides a foundational standpoint on how the concept of learning takes place and how the teaching process influences its development. The behavioural paradigm assumes that if learners are willing and motivated and their teachers communicate clearly and effectively, then adequate learning can take place. This view also considers the impact of external stimuli on the effectiveness of learning among students. Based on the role of the environment in enhancing learning in this paradigm, educators are expected to ensure a learning-friendly environment to benefit learners. The behavioural paradigm also places a huge emphasis on learning via repetition and conditioning while focusing less on mental activities. Behaviourists insist on repeating a behavioural pattern until it becomes automatic, hence this paradigm does not explain the cognitive aspect (Arnold and Yeomans, 2005).

While the behavioural paradigm places enormous emphasis on the learner's external behaviour, the cognitive paradigm highlights knowledge acquisition and the mental assimilation (cognitive) process. The view also ascertains how information is received, processed, and assimilated by the mind. While these internal and psychological processes are reasonably complex, understanding them provides a solid step as to how effective learning can occur (Schoenfeld, 1987). Moreover, learning according to the cognitive paradigm goes beyond the learner's actions to how they know what they know (Jonassen, 1991). In addition, the learner is actively pivotal to this learning process as the individual accesses, transfers, and assimilates new information (Wilson and Peterson, 2006). Shuell (1986) asserts that the cognitive paradigm has significantly influenced the notion of learning theory, describing it as an active and higher level learning process. Furthermore, the cognitive approach of learning highlights the correlation between stimuli and responses, observing learning as a sequence of overlying stages of information processing and rendering (Chalkley, 1996). This implies that learning under the cognitive paradigm is principally the concept of receiving and processing information from the environment.

Constructivism

The shift to constructivism has triggered a sudden interest in the various ways by which an individual perceives the world at large with reference to knowledge, experiences, and interaction with peers and team groups. While two different active individuals may be exposed to the same thread and pattern of information, they are expected to develop their ability to understand differently. This occurs because a learner develops varying levels of cognitive conflict (anxiety and pressure) that is a fundamental element of the learning process (Rowell and Dawson, 1979). In the constructivist paradigm, the learner conceptualizes innovative concepts based on several factors, including motivation levels, existing knowledge, and interrelationship with other people. However, in the behavioural paradigm, although a learner is reserved, there is no guarantee that his or her psyche is effectively occupied with learning. Therefore, learning according to the constructivist view posits the concept of learning as a thoughtful process of constructing understanding by applying new information to that already existing to create a coherent idea.

This understanding can be constructed in various aspects, depending on the learner's motivation level, experience, and learning styles (Earl and Katz, 2006). According to Hoover (1996), four key assertions underpin the constructivist paradigm. Firstly, the learning process which in this view is active; secondly, educators helping learners by providing them with meaningful activities; thirdly, sufficient time offered to learners to help them assimilate their experiences and learning process; and finally, learners interacting effectively with their social environment to ensure a holistic learning process. The activeness of learners in the constructivism paradigm increases their independence, depth of learning, satisfaction, thinking skills, and retention. According to Confrey and Kazak (2006), there are two critical and effective types of constructivism: social constructivism and cognitive constructivism. The social aspect (Vygotsky's constructivism) suggests that knowledge construction relates to learners' usage of the language as well as social interaction and encounters. Simply put, educators in this aspect are to be conversant with the various social and cultural impacts on learning. On the other hand, the cognitive or developmental aspect (Piaget's constructivism) suggests that knowledge construction relates to learners' active participation and experience with their immediate environment (Confrey and Kazak, 2006). Therefore, educators in this aspect are to be conversant with the various levels of cognitive development and be able to administer their lessons according to the students' ability and cognitions at various levels. Constructivism, as a learning approach, offers students the opportunity to be problem-solvers in complex work environments (Crawford and Jenkins, 2018).

Neurophysiological theories

According to Hergenhahn (1988), the different neurophysiological aspects involved in thinking, learning, observing, and assimilating work in tandem with the roles of the different hemispheres of the brain. The neurophysiological theory posits that the brain hemisphere process information in two different ways. While the right hemisphere processes the impulsive and intuitive information, the left hemisphere deals with the numerical and sequential aspects of information. The neurophysiological theory further believes that conventional education tends to emphasize the intelligence most common with the left hemisphere. However, as both aspects do not work in isolation, it is a daunting task to design an educational experience that would cater for one hemisphere. Hence, this limitation prevents the holistic adoption of this theory when possible methods of fostering certain learning aspects are required.

Humanist theories

There is a growing perception that learning is a function of personal growth that is fostered by interpersonal connections. The humanist theory believes that students themselves can achieve significant learning and personal development. In this case, the learner is central to the learning process, making the teacher's role to be best described as a facilitator. The humanist paradigm views individuals' as having control over their decisions and destiny by virtue of career paths, choices, and goals. Hence, educators are required to ensure a conducive and positive educative environment to enable students to thrive emotionally, build self-confidence, and flourish holistically.

One key aspect of the humanist approach is the concept of experiential learning. This concept deals with learning from one's own experience and events to improve the current learning stance (Chalkley, 1996). The views of Karl Rogers and Abraham Maslow are further key in elucidating the underpinnings of the humanistic theories. According to Rogers (1983), the learning process must take into cognizance the cognitive and affective aspects to ensure a holistically trained and independent individual. Rogers also posits that the learning process under the humanism paradigm implies that the learning process is conducted and appraised by the learner in question. On the other hand, Maslow places physiological needs at the lowest and self-actualization at the highest level in his hierarchy of motivation. Maslow opines that individuals can only attain the next level if they successfully navigate and address their lower level needs (Maslow, 1968). Albert Bandura, often regarded as the father of the humanism paradigm (behavioural psychology), used the term 'observational learning' to further elucidate this type of learning.

Both the neurophysiological and humanist's intricacies go beyond the scope of this study but have been mentioned to gain an all-inclusive grasp of the learning scope. From the behaviourism theory to the humanism view, a knowledge continuum exists along which the scope of learning can be further understood. At one end, learning is characterized by the educators who dictate the content and learning method, whereas at the other end, the learners dictate the learning process themselves. In reality, various educational curricula and educators adopt and combine some pedagogical approaches depending on what they plan to achieve. These approaches are discussed next.

Developing employability skills

With learning and learning outcomes understood, it is germane to understand how employability skills can be developed among students. This can be achieved in universities by supporting its educators and facilitators with the relevant resources and conditions, improving curriculum and course design, and exposing students to real-work activities and career services amongst others. Yorke and Knight (2006) have discussed the concept of employability exhaustively over time. They define the concept as a collection of skills, knowledge, and personal dispositions that make a person more desirable to employers and function effectively when employed. They posit that employability needs to be construed as a learning outcome that offers great benefits to the lifelong learning process and life in general.

According to various researchers, the role of universities in developing their learners to achieve industry readiness depends on their innovative and inventive educational approaches. This is because learners need more than discipline-specific skills (academic skills) alone to be employable upon graduation (Nagarajan and Edwards, 2015). In developing learners, the concept of teaching is important. According to Kramer (1999), teaching, like other professions, is a professional, organized, and intentional activity that comprises its ability, skill, and knowledge attained through regular practice, training, and development. Several researchers have offered a variety of different teaching approaches by means of which learners can learn effectively. However, in fostering student participation and knowledge assimilation, three approaches are discussed as posited by Steenekamp (2004). They include dialogic, problem-solving, and reflective teaching approaches.

Dialogic approach

Several researchers have highlighted the significant role of dialogue in teaching and learning in ensuring effective and meaningful learning for students. 'Dialogue' as a term has its meaning traced to the Greek words *dia* and *logos*, with dia meaning 'two' as well as 'across', indicating that the concept can be applicable beyond two; and *logos* meaning 'reason or thought' (Gravett, 1996; Gravett, 2001). Therefore, from both individual meanings, dialogue refers to thoughts, opinions, views, judgements, and knowledge that can occur among several people. In an educational scenario, this denotes that educators and students or students amongst themselves are required to convey their thoughts, opinions, views, judgements, and knowledge during a learning period. With reference to this teaching approach, educators are expected to oversee the learning process of students, ensuring that they communicate their thoughts, opinions, views, judgements, and knowledge. Moreover, this teaching process affords both educators and learners the opportunity to forge a communicative didactic relationship, which is made possible as a result of the mutual sharing of one another's thoughts, opinions, views, judgements, and knowledge.

Problem-solving approach

The problem-solving teaching approach is a strategy employed by educators to ensure their students learn effectively and meaningfully. Educators can achieve this by formulating thought-provoking problem situations that ensure that learners are cognitively engaged in proffering solutions to designed questions. Therefore, this approach ensures that students are holistically developed and possess higher order thinking skills to function effectively in real-life situations (Gravett, 1996, 2001).

Reflective teaching approach

The reflective teaching approach is employed by educators to assess the learning curve and process of learners. The need for educators to reflect on what they are teaching as well as on educational outcomes is critical to the holistic development of their students. This strategy can help shape the mentality of educators and offer them reasons to revisit and revamp their training methods. One way by which educators can conduct self-assessment is through the collection of feedback from students. Biggs (1999) describes self-assessment as a critical tool that differentiates expert teachers from non-experts as the willingness to collect student feedback indicates the willingness to improve on their teaching methods. However, in addition to student feedback, educators are to integrate self-assessment and self-reflection in drawing up a holistic report of their performance to ensure an all-inclusive teaching approach.

Pedagogical approaches in developing employability skills

Final-year research projects

Several researchers have discussed the various benefits of final-year research or semester projects and its contribution to graduate employability. These research projects provide an opportunity to test the individual's learning integrity and experience via an exploration of facts that culminate in the knowledge gathered through various academic levels. Some of these research projects assist the students in establishing connections between their chosen careers and the world of work. These research activities also promote holistic thinking among students, increase their self-confidence and motivation, enrich their academic understanding of their chosen discipline, and promote the development of key skills (problem-solving, critical thinking, organizational and interpersonal skills) (Kift et al., 2013; Kinash et al., 2014).

Career advice and development

In improving the employability of learners and their work profile, the role of career management and development cannot be overemphasized (Bridgstock, 2009). According to Kuijpers and Scheerens (2006), career advice and development help prepare learners for industry by developing their job-searching skills such as interview preparation, curriculum vitae (CV) design, self-reflection and self-assessment abilities, and networking competencies. Doyle (2012) suggests that career development involves several activities that cater to the career needs of learners, such as professional workshops and seminars.

Extracurricular activities

Several studies have been conducted in examining the role of extracurricular activities in enhancing the employability of learners. This is accomplished by the combination of course work and experiential learning with the external community (Parker *et al.*, 2009). According to several researchers, extracurricular activities are activities that complement conventional classroom events that occur after schooling hours. Terms used to describe such activities include learners' activities, socializing, co-curricular and extra-class activities (Esa *et al.*, 2012). The formal and informal activities which comprise extracurricular activities make it possible for students who engage in such activities to develop soft skills (teamwork, problem-solving, communication, managerial and leadership skills) and abilities which industry employers constantly seek (Shakir, 2009). Activities such as citizenship development, workshops, conferences, community-based service learning, and award issuance based on activity organization and involvement, amongst others, are all various forms of extracurricular activities (Kinash *et al.*, 2014). Students who also participate in club and societal activities possess certain advantages when seeking jobs after graduation, thereby improving their employability. Consequently, Nemanick and Clark (2002) suggest that the presence of extracurricular activities on a job seeker's résumé improves the chances of employment. Apart from displaying less risky behaviours, students who participate in extracurricular activities usually exhibit high academic competencies (Lau *et al.*, 2014) and develop better career skills (Shiah *et al.*, 2013) and improved connectivity with the community (Wilson, 2009). Participation in extracurricular activities further helps students with supportive relationships or mentorships that can provide constructive feedback during the learning process (Wilson, 2009).

International exchange programmes

With the predominant wave of globalization and internationalization, the need for learners to operate in diverse cultural contexts cannot be over-emphasized. This has prompted several universities globally to incorporate global components into their respective vision and mission statements. One of the ways through which universities can ensure this is the fostering of international exchange programmes that can improve learners' overall learning, cognitive abilities, professional development, cultural sensitivity, and ultimately, employability. These employability development programmes facilitate the attainment of valuable experience and expertise, enabling learners to function as global citizens who are adequately equipped to function in the environment in which they find themselves (Crossman and Clarke, 2010). Therefore, graduates who take part in these global exchange programmes may likely be more employable than those with solely local exposure. Furthermore, international exchange programmes provide learners with ample opportunity to develop their competencies and skills, network with their peers of different orientations and backgrounds, develop new language skills, and attain valuable work experience (Crossman and Clarke, 2010; Kinash *et al.*, 2014).

Mentoring opportunities

In improving the employability of learners, another key strategy is the encouraging of mentoring among students. Mentoring is a social learning and interactive opportunity that improves learners' transition from a lecture-room setting to a work setting via industry involvement. According to Smith-Ruig (2014), industry mentors provide students with the knowledge required to thrive in their chosen careers and to succeed in the world of work. Moreover, the presence of industry mentorship encourages career development and outcomes among students, hence employability improvement is guaranteed. Levesque *et al.* (2005) also suggest that industry mentoring provides students with career information, increased commitment, political and material support, on-the-job training, motivation, workplace understanding and realities, facilitation of connections for students, and role modelling.

Visits to industry events

In further improving the employability of learners, the benefits of attending industry events cannot be overestimated. Like the role of mentoring, attending industry events and networking provides students with the opportunity to establish rapport with industry employers, thereby improving their learning process and ultimately employability (Watanabe, 2004). This can be achieved by establishing forums, symposiums, and workshops (online or face-to-face meetings) comprising learners, alumni, faculty members, and industry stakeholders for interaction purposes, continuous learning, and knowledge exchange.

Part-time employment for students

Another key strategy in improving the employability of learners is students' involvement in part-time employment (work activities) while studying. According to Smith (2009), students who work and learn full-time are afforded the prospects of transitioning into their chosen careers with relative ease as opposed to those who do not. Moreover, part-time employment opportunities provide learners with key skills and generic competencies such as communication skills, teamwork skills, problem-solving skills, and critical thinking abilities.

Work-integrated learning and placement opportunities

Work-integrated learning (WIL) is another strategy by means of which the employability of learners is developed. This learning approach provides learners with the opportunity to understand and holistically apply the knowledge obtained from the lecture room to the reality of their work environment. These work activities provide learners with workplace skills and abilities which are critical to their employability status. Work placements provide practical opportunities that help ease the transition from lecture-room activities to real-work activities among students, helping them in the following ways: improved learning process, improved educational performance and outcome, lifelong and experiential learning, communicative abilities, critical skills, increased learning processes, analytical thinking, organizational abilities, and improved employability outcomes (Kinash *et al.*, 2014).

Several studies also found that students who undergo WIL tend to possess job knowledge, self-confidence, work-ready attitudes, motivation, and a critical understanding of workplace culture and expectations. It also helps learners to understand that career success is dependent not only on academic success but also on understanding their chosen career, an understanding which is obtainable via WIL. Other studies have shown that WIL provides students with a seamless transition from the lecture room to the world of work, thus enhancing career development after graduation (Jackson *et al.*, 2015). Clarke (2018) suggested that for WIL to be effective in delivering these various benefits, its activities must be relevant, and purposefully incorporated and aligned with the curriculum of the university. Over time, a wide range of terms have been used to represent WIL. They include work experience, simulations, volunteering, fieldwork, work-based learning, project-based learning, project-based work, co-operative education, internship, and service-learning (McLennan and Keating, 2008). Therefore, WIL is a broad term used to describe educational programmes where work experience is incorporated into the university curriculum to complement the lecture-room experience and beyond.

Portfolio development

Another way by means of which employability skills can be developed among students is via the development of graduate portfolios, records of accomplishment, and success profiles. These documents represent a collection of skills and competencies which indicate the employability level among learners (Kinash *et al.*, 2014). They also constitute documents that highlight learners' learning outcomes and their corresponding employability implications. Learners who develop portfolios are fully aware of their employability profile, strengths, and weaknesses of their chosen careers. Portfolios also provide universities with an update on learners and are aware of possible aspects along which to channel their efforts (Oliver and Whelan, 2011).

Volunteering and community engagement

In further improving the employability of learners, engaging with the immediate community plays a critical role in overall growth and fosters a holistic transformation experience. Likewise, volunteering opportunities are pivotal in developing several skills and competencies among learners such as teamwork abilities, leadership skills, and personal qualities (Parker *et al.*, 2009). According to Hall *et al.* (2009), volunteering opportunities provide students with adequate work experience, academic development, and a sense of civic responsibility.

Project-based learning

Generally, the conventional lecture room approach helps in disseminating educational information to students. In further enhancing graduate attributes and competencies, several learner-centred (student-centred) approaches are significant, including problem-based and project-based learning, inquiry learning, discovery learning, and case-based teaching approaches. They are student-centred because learners are primarily responsible for their learning rather than the conventional lecture-based approach in disseminating educational information (Prince and Felder, 2006). Project-based learning (PBL) is one of the key mechanisms that help students to develop skills and attributes via several learning experiences. This education model engages students in the learning process through a carefully designed real-life project that helps in the development of life-enhancing skills and essential knowledge (Moalosi *et al.*, 2012). Project tasks are characterized by thought-provoking and investigative activities that provide learners with the chance to work independently or in teams over a period. Apart from making learning more exciting for students by engaging them, this approach stimulates their learning interest, hence fostering several skills, including critical thinking, communicative, higher level thinking, teamwork, and problem-solving skills (Prince and Felder, 2006).

Student government participation

Another key strategy by means of which students can develop their employability skills is through involvement in well-structured government activities by assuming leadership positions while on campus. According to Alviento (2018), participating in student government activities enhances continuous learning, career competence, personal growth and development, self-esteem and confidence, decision-making skills, leadership skills, problem-solving skills, and civic engagement. Moreover, students who engage in leadership positions during their university days are more attractive in the eyes of employers because of their employability value and marketability (Gacutan, 2006). Furthermore, participation in student government activities provides learners with the ability to regulate their thoughts and emotions, leading to improved self-confidence.

Field trips

In further developing employability skills among students, the integration of field trips or study excursions as part of the university curriculum plays a significant role. This pedagogical method involves sponsoring students to visit certain places with the intent to acquire learning experience under the supervision of their teachers (DeWitt and Storksdieck, 2008). Field trips are outings and excursions that provide enhanced learning for students as well as increased practical knowledge of their subject area (Bamberger and Tal, 2008).

Simulation and role-play

Both simulation and role-plays share several characteristics, functions, and features and their role in developing employability skills are similar. Simulation deals with the behavioural imitation of a situation or process that mirrors reality via something suitably analogous. Simulation is also a problem-solving activity that involves the personalities and opinions of students involved. While simulation deals with students behaviours in an activity or a process, role-play involves portraying a real-life character in achieving the required knowledge (Pierce and Middendorf, 2008). Moreover, simulation is time-consuming, complex, and somewhat rigid, whereas role-play is usually short term, easily structured, and flexible. This feature portrays simulation as a broader concept when compared to role-playing. As an effecting teaching method, role-playing ensures students take on characters known in their day-to-day activities (projection in real situations), whereas in simulations they deal with real-life situations (Qing, 2011; Bhattacharjee and Ghosh, 2013). Simulation and role-playing further provide students with improved motivation and creativity, increased confidence, reduced tension and anxiety, improved communication skills, deeper understanding of the subject matter, improved academic knowledge, self-reflection, improved decision-making, team-building skills, and identification of strengths and weaknesses (Shapiro and Leopold, 2012; Krebt, 2017). Unlike the traditional teaching method in which students are passive, this pedagogical approach ensures the activeness of students (speaking, acting, and sharing) in the learning process (Altun, 2015).

Integration of technical competitions

In developing graduate employability, the integration of technical competitions into the university curricula cannot be overstated. As a pedagogical teaching approach, this thought-provoking strategy involves students in the learning process and can exist in three forms. Students can work individually (individualistic method), compete with each other (competitive method), or work with each other (collaborative method) (Regueras *et al.*, 2011). They go on to suggest that these three approaches provide a high level of performance, motivation, and engagement with the body of knowledge. Moreover, the competitive approach improves subject-knowledge understanding, increases satisfaction, captivates the interest, and increases participation and the critical thinking abilities of students (Ahlgren and Verner, 2013). Kundu and Fowler (2009) also suggest that technical competitions provide students with opportunities to develop various skills and competencies, including teamwork skills, communication skills, interpersonal competencies, and problem-solving abilities. Technical competitions further provide students with deeper learning tendencies, innovative abilities, and real-world opportunities to address industry problems (Battisti *et al.*, 2011; Fingerut *et al.*, 2013).

Engagement in sports activities

Students' engagement in sports activities and competitions is also beneficial to their employability development. Apart from contributing to their cognitive skills, sports competitions provide students with a healthy lifestyle, good physical and mental health, reduced risky behaviours, and improved academic performance (Bailey *et al.*, 2013). For students to achieve these outcomes, integrating physical education activities (PEA) as part of the overall educational process is pivotal (Telford *et al.*, 2012). Moreover, active participation during sporting activities enhances behavioural traits such as effective communication skills, motivation, and self-confidence (Jonker *et al.*, 2010). Similarly, participation in sports activities leads to increased productivity, reduced stress due to blood flow to the brain, mental alertness, and improved mood, thereby improving academic output.

Assessing employability skills

Having discussed various teaching approaches in developing employability skills, it is pertinent to discuss assessment as a critical component of the learning process. Rogler (2014) highlights the role of assessment in teaching and learning and how it provides an opportunity to check the influence of educators and faculty staff in developing learners. Ndalichako (2015) suggests that assessment provides many opportunities to monitor and improve the learning process and also provides valuable feedback for learners so that learning errors and difficulties can be mitigated. It is also the process of collecting, recording, and interpreting information on learners' progress and educational outcomes and ways to foster their growth. Assessment also provides learners with an idea of their weaknesses and ways to address them to gain required learning outcomes, hence improving their learning process. It goes a long way in improving motivation among learners and enabling them to establish good study habits. Apart from learners' benefit, assessment also assists educators in supporting learners' progress and regulating their teaching methods when necessary to maintain and achieve intended learning and teaching outcomes. Therefore, assessment is a critical feature of both teaching and learning as it plays a part in ensuring the quality of educational activities. Agrey (2004) views assessment as a critical tool in student engagement that aims to retain their commitment and attitude towards learning. These previously mentioned influences of assessment on teaching and learning outcomes have prompted several academics and investigators to motivate the inclusion of assessment into the teaching and learning process. Consequently, the role of educators in assessment and ensuring it leads to effective teaching outcomes is extremely critical. Apart from facilitating the training of learners, educators are also tasked with lecture-room assessment in ways and avenues to enhance their intended learning outcome.

Furthermore, several researchers have provided various purposes of assessment. According to Kellough and Kellough (1999), there are seven key functions of assessment: fostering students' learning, ascertaining learners' strengths and weaknesses, evaluating the efficiency of teaching strategies, evaluating the efficiency of curricula, evaluating the efficiency of educational activities, assisting in decision-making for both learners and educators, and providing parents with information with regard to the success level of their children (Kellough and Kellough, 1999). Likewise, Loubser (1993) provides four functions of assessment: monitoring of learners' progress, ascertaining learners' shortcomings, ascertaining learners' strengths, and examining the effectiveness of teaching and learning. However, Sebate (2012) suggests six roles of assessment. They include screening (identifying learners who require special attention), diagnosis (ascertaining learners' strengths and weaknesses), record keeping (keeping track of learners' progress), feedback (generating information on both learner and educator success), certification (providing learners with the necessary documentation indicating the completion of a competence level), and selection (providing learners with necessary ideas about their career choices and further studies). Four purposes of assessment (diagnostic, formative, summative, and performance and portfolio assessment) are discussed next.

Diagnostic assessment

Apart from measuring and evaluating the existing knowledge and capabilities of learners, the diagnostic assessment provides educators with a holistic understanding of those existing knowledge and capabilities brought by learners to the learning space (Booyse and Du Plessis, 2014). In this form of assessment, evaluating learners' existing knowledge usually takes place at the commencement of a level (grade) to examine the previous knowledge of learners. This outcome-based assessment helps to diagnose the weaknesses and struggles faced by learners, hence providing educators with an idea of ways to mitigate these problems (Van der Merwe, 2011). Similarly, Sebate (2012) posits that this form of assessment is significant in examining the learners' strengths and challenges in order to enhance the strengths and rectify possible problems. Hence, the diagnostic assessment provides educators with various teaching and pedagogical strategies (formal and informal) to enhance the employability skills of learners.

Formative assessment

A distinct feature of this form of assessment is that it occurs just before or during a lecture or teaching experience, intending to generate proof that can be utilized in learning improvement (Killen, 2015). Likewise, Booyse and Du Plessis (2014) assert that assessment becomes formative when the generated proof is utilized effectively to improve teaching methods and approaches in developing learners. Moreover, the formative assessment occurs through various approaches (formal and informal), hence it aims at improving the teaching process, thereby benefiting learners. Its role in improving and developing the learning process has seen formative assessment also being regarded as assessment for learning. In addition, assessment for learning is significant in developing learners and informing them of their possible strengths and weaknesses. The informal aspect of formative assessment helps to monitor the progress of learners and this can be achieved via teacher–learner exchanges and teacher-observations. This form of assessment is largely associated with the constructivism paradigm previously discussed because learners are actively participative in the learning process.

Flórez and Sammons (2013) further noted that formative assessment is significant in the feedback process for both educators and learners, hence teaching and learning activities are enhanced and improved. Heritage (2007) proposed four key components of formative assessment. They are gap identification (teaching and learning), a feedback mechanism (learners and educators responses), student involvement (allowing learners to participate in the learning process), and learning progression (allowing learners to flourish in the learning process) (Heritage, 2007). Moreover, the adoption of formative assessment in the lecture room requires educators to possess these four specific elements: domain knowledge (understanding concepts and knowledge within a discipline), pedagogical content knowledge (understanding various methods of teaching), learners' previous learning (understanding previous knowledge and level of educational content attained by their learners), and assessment knowledge (understanding the quality of assessment to evaluate learners) (Heritage, 2007).

Summative assessment

A distinct feature of this form of assessment is that it is more effective on a large scale that focuses on the most critical aspects and outcomes of learning. According to Dixon-Román (2011), summative assessments help in post hoc decisions such as evaluation, reports, promotions, certification decisions, and accountability discussions. Similarly, Booyse and Du Plessis (2014) assert that summative assessments decide the productivity and success of learners, usually at the end of a learning or teaching process. Hence, this form of assessment provides learners with accumulative grades that contribute to the overall rankings or evaluations after a specific period. This form of assessment further provides the teachers and even parents with information of learners' progress and identifies room for possible improvement. The users of this form of assessment include learners themselves, parents, employers, and even society, if possible. While the formative assessment fosters learning, the summative tends towards the monitoring of learning.

Performance and portfolio assessment

This type of assessment presents learners with hands-on activities or tasks requiring them to apply their skills and knowledge acquired from the lecture room in proffering solutions (Nitko, 2004). The performance assessment also adopts a well-defined benchmark to evaluate the quality of effort put in to achieve the outcome required. Moreover, this assessment type provides a clear opportunity for learners to display their retentive memory, understanding ability, and analytical and organizational skills in achieving an expected outcome. On the other hand, portfolio assessment is achieved by taking inventory of learners' work collections. A portfolio goes beyond a folder containing several progress reports, writing materials, and written observations. It is a well-documented collection of tasks undertaken and completed, deliverables, and achievements over a period (Orlich *et al.*, 2012). A well-document portfolio can encourage independence and motivation among learners, hence engaging them in self-reflection.

Summary

This chapter reviewed the concept of learning as well as several ways to assess learning outcomes on students. Different learning theories were discussed for a robust understanding of the intricacies behind the concept of learning. The chapter also discussed teaching approaches and examined various teaching methods by means of which higher education can develop employability skills. This was necessary because to develop an employability improvement framework for graduates, there is an urgent need to be aware of the various ways by means of which educators can further revisit and strengthen their curriculum to produce skilled graduates for the future. Furthermore, the chapter discussed the role of assessment in teaching and learning and how it provides an opportunity to monitor the influence of teachers and faculty staff in learning development. The next chapter reviews the various employability and generic competency frameworks in several international contexts. The chapter also reviews several international research studies conducted on employability and the employability skills required for graduate success.

References

Agrey, L. 2004. The pressure cooker in education: Standardized assessment and high stakes. *Canadian Social Studies*, 38(3), p. 3.

Ahlgren, D.J. and Verner, I.M. 2013. Socially responsible engineering education through assistive robotics projects: The robowaiter competition. *International Journal of Social Robotics*, 5(1), pp. 127–138.

Altun, M. 2015. Using role-play activities to develop speaking skills: A case study in the language classroom. *International Journal of Social Sciences and Educational Studies*, 1(4), pp. 27–33.

Alviento, S. 2018. Effectiveness of the performance of the student government of North Luzon Philippines State College. *Research in Pedagogy*, 8(1), pp. 1–16.

Arnold, C. and Yeomans, J. 2005. *Psychology for teaching assistants*. Sterling: Trentham Books.

Bailey, R., Hillman, C., Arent, S. and Petitpas, A. 2013. Physical activity: An underestimated investment in human capital? *Journal of Physical Activity and Health*, 10(3), pp. 289–308.

Bamberger, Y. and Tal, T. 2008. Multiple outcomes of class visits to natural history museums: The students' view. *Journal of Science Education and Technology*, 17(3), pp. 274–284.

Battisti, F., Boato, G., Carli, M. and Neri, A. 2011. Teaching multimedia data protection through an international online competition. *IEEE Transactions on Education*, 54(3), pp. 381–386.

Bhattacharjee, S. and Ghosh, S. 2013. Usefulness of role-playing teaching in construction education: A systematic review. *Proceedings of the 49th ASC Annual International Conference*, San Luis Obispo, CA.

Biggs, J. 1999. *Teaching for quality learning at university*. Suffolk: Society for Research into Higher Education (SRHE) & Open University Press.

Booyse, C. and Du Plessis, E. 2014. *Curriculum studies: Development, interpretation, plan and practice*. Pretoria: Van Schaik, University of South Africa.

Bridgstock, R. 2009. The graduate attributes we've overlooked: Enhancing graduate employability through career management skills. *Higher Education Research & Development*, 28(1), pp. 31–44. doi: 10.1080/07294360802444347.

Chalkley, S.T. 1996. *Student-centred quality improvement systems in manufacturing engineering higher education* (Doctoral dissertation, Brunel University School of Engineering and Design, Ph.D. thesis, London).

Clarke, M. 2018. Rethinking graduate employability: The role of capital, individual attributes and context. *Studies in Higher Education*, 43(11), pp. 1923–1937.

Confrey, J. and Kazak, S. 2006. A thirty-year reflection on constructivism in mathematics education in PME. In *Handbook of research on the psychology of mathematics education: Past, present and future* (pp. 305–345). Leiden: Brill. E-Book ISBN: 9789087901127.

Crawford, R. and Jenkins, L.E. 2018. Making pedagogy tangible: Developing skills and knowledge using a team teaching and blended learning approach. *Australian Journal of Teacher Education*, 43(1), p. 8.

Crossman, J.E. and Clarke, M. 2010. International experience and graduate employability: Stakeholder perceptions on the connection. *Higher Education*, 59(5), pp. 599–613.

DeWitt, J. and Storksdieck, M. 2008. A short review of school field trips: Key findings from the past and implications for the future. *Visitor Studies*, 11(2), pp. 181–197.

Dixon-Román, E. 2011. Assessment, teaching, and learning. *The Gordon Commission on the Future of Assessment in Education*, 1(2), pp. 1–8.

Doyle, D.E. 2012. Developing occupational programs: A case study of community colleges. *The Journal of Technology Studies*, 38(1), pp. 53–62.

Earl, L. and Katz, S. 2006. *Rethinking classroom assessment with purpose in mind*. Retrieved from: http://files.eric.ed.gov/fulltext/EJ780818.pdf.

Ertmer, P.A. and Newby, T.J. 2013. Behaviourism, cognitivism, constructivism: Comparing critical features from an instructional design perspective. *Performance Improvement Quarterly*, 26(2), pp. 43–71.

Esa, A., Md Yunus, J. and Kaprawi, N. 2012, May. The implementation of the generic skills through cocurricular at the polytechnics to fulfil the industrial needs in Malaysia. *Proceedings of the International Conference on Education*, Athens.

Fingerut, J., Orbe, K., Flynn, D. and Habdas, P. 2013. Focusing on the hard parts: A biomechanics laboratory exercise. *Bioscene: Journal of College Biology Teaching*, 39(1), pp. 10–15.

Flórez, M.T. and Sammons, P. 2013. *Assessment for learning: Effects and impact*. Reading, MA: CfBT Education Trust.

Gacutan, C.G. 2006. *The level of participation of the SGO officers in the school development of Candon National High School* (Undergraduate Thesis, CCC University of Northern

Philippines-Candon City, Ilocos Sur, Bachelor of Arts Major in Political Science-History, Philippines).

Gravett, S.J. 1996. The assessment of learning in higher education: Guiding principles. *South African Journal for Higher Education*, 1, pp. 76–81.

Gravett, S.J. 2001. *Adult learning: Designing and implementing learning events: A dialogical approach.* Pretoria: Van Schaik Publishers.

Hall, M., Lasby, D., Ayer, S. and Gibbons, W. 2009. *Caring Canadians, involved Canadians: Highlights from the 2007 Canada survey of giving, volunteering and participating.* Ottawa, ON: Statistics Canada.

Hergenhahn, B.R. 1988. *An introduction to theories of learning* (3rd ed.). Upper Saddle River, NJ: Prentice-Hall International.

Heritage, M. 2007. Formative assessment: What do teachers need to know and do? *Phi Delta Kappan*, 89(2), pp. 140–145.

Hoover, W.A. 1996. The practice implications of constructivism. *SEDL Letter*, 9(3), pp. 1–2.

Jackson, D., Ferns, S., Rowbottom, D. and McLaren, D. 2015. *Working together to achieve better work integrated learning outcomes: Improving productivity through better employer involvement.* Adelaide: ACEN.

Jonassen, D.H. 1991. Objectivism versus constructivism: Do we need a new philosophical paradigm? *Educational Technology Research and Development*, 39(3), pp. 5–14.

Jonker, L., Elferink-Gemser, M.T., Toering, T.T., Lyons, J. and Visscher, C. 2010. Academic performance and self-regulatory skills in elite youth soccer players. *Journal of Sports Sciences*, 28(14), pp. 1605–1614.

Kellough, R.D. and Kellough, N.G. 1999. *Secondary school teaching: A guide to methods and resources: Planning for competence.* Englewood Cliffs, NJ: Prentice-Hall.

Kift, S., Butler, D., Field, F., McNamara, J., Brown, C. and Treloar, C. 2013. *Curriculum renewal in legal education* (Final report). Sydney: Office for Learning and Teaching. Retrieved from: www.olt.gov.au/system/files/resources/PP9-1374_Kift_Report_2013_1.pdf.

Killen, R. 2015. *Teaching strategies for quality teaching and learning.* Cape Town: Juta and Company.

Kimble, G.A. 1964. Categories of learning and the problem of definition: Comments on Professor Grant's paper. In *Categories of human learning* (pp. 32–45). Cambridge, MA: Academic Press.

Kinash, S., Crane, L., Knight, C., Dowling, D., Mitchell, K., McLean, M. and Schulz, M. 2014. Global graduate employability research. *A report to the Business20 Human Capital Taskforce (Draft)*, Australia.

Kolb, D.A. 1984. *Experiential learning: Experience as the source of learning and development.* Englewood Cliffs, NJ: Prentice-Hall.

Kramer, D. 1999. *OBE teaching toolbox.* Florida Hills: Vivlia Publishers.

Krebt, D.M. 2017. The effectiveness of role play techniques in teaching speaking for EFL college students. *Journal of Language Teaching and Research*, 8(5), pp. 863–870.

Kuijpers, M.A.C.T. and Scheerens, J. 2006. Career competencies for the modern career. *Journal of Career Development*, 32(4), pp. 303–319.

Kundu, S. and Fowler, M.W. 2009. Use of engineering design competitions for undergraduate and capstone projects. *Chemical Engineering Education*, 43(2), pp. 131–136.

Lau, H.H., Hsu, H.Y., Acosta, S. and Hsu, T.L. 2014. Impact of participation in extracurricular activities during college on graduate employability: An empirical study of graduates of Taiwanese business schools. *Educational Studies*, 40(1), pp. 26–47.

Levesque, L.L., O'Neill, R.M., Nelson, T. and Dumas, C. 2005. Sex differences in the perceived importance of mentoring functions. *Career Development International*, 10(6/7), pp. 429–443.

Loubser, C.P. 1993. Evaluation. In W.J. Fraser, C.P. Loubser and M.P. Van Rooy (Eds.), *Didactics for the undergraduate student* (2nd ed., pp. 187-207). Durban: Butterworths.

Maslow, A. 1968. *Towards a psychology of being* (2nd ed.). New York: Van Nostrand.

McLennan, B. and Keating, S. 2008, June. Work-integrated learning (WIL) in Australian universities: The challenges of mainstreaming WIL. *Proceedings of the ALTC NAGCAS National Symposium*, Melbourne, Australia, pp. 2–14.

Moalosi, R., Oladiran, M.T. and Uziak, J. 2012. Students' perspective on the attainment of graduate attributes through a design project. *Global Journal of Engineering Education*, 14(1), pp. 40–46.

Nagarajan, S. and Edwards, J. 2015. The role of universities, employers, graduates and professional associations in the development of professional skills of new graduates. *Journal of Perspectives in Applied Academic Practice*, 3(2).

Ndalichako, J.L. 2015. Secondary school teachers' perceptions of assessment. *International Journal of Information and Education Technology*, 5(5), pp. 326–330.

Nemanick, R.C., Jr. and Clark, E.M. 2002. The differential effects of extracurricular activities on attributions in resume evaluation. *International Journal of Selection and Assessment*, 10(3), pp. 206–217.

Nitko, A.J. 2004. *Educational assessment of students* (4th ed.). Upper Saddle River, NJ: Merrill.

Oliver, B. and Whelan, B. 2011. Designing an e-portfolio for assurance of learning focusing on adoptability and learning analytics. *Australasian Journal of Educational Technology*, 27(6), pp. 1026–1041.

Orlich, D.C., Harder, R.J., Callahan, R.C., Trevisan, M.S. and Brown, A.H. 2012. *Teaching strategies: A guide to effective instruction*. Boston, MA: Cengage Learning.

Parker, E.A., Myers, N., Higgins, H.C., Oddsson, T., Price, M. and Gould, T. 2009. More than experiential learning or volunteering: A case study of community service learning within the Australian context. *Higher Education Research & Development*, 28(6), pp. 585–596.

Pierce, D. and Middendorf, J. 2008. Evaluating the effectiveness of role playing in the sport management curriculum. *International Journal of Sport Management and Marketing*, 4(2), pp. 277–294.

Popham, W.J. 2008. *Classroom assessment: What teachers need to know* (5th ed.). Boston, MA: Pearson Education.

Prince, M.J. and Felder, R.M. 2006. Inductive teaching and learning methods: Definitions, comparisons, and research bases. *Journal of Engineering Education*, 95(2), pp. 123–138.

Qing, X.U. 2011. Role-play: An effective approach to developing overall communicative competence. *Cross-Cultural Communication*, 7(4), pp. 36–39.

Regueras, L.M., Verdu, E., Verdu, M.J. and De Castro, J.P. 2011. Design of a competitive and collaborative learning strategy in a communication networks course. *IEEE Transactions on Education*, 54(2), pp. 302–307.

Rogers, C.R. 1983. *Freedom to learn for the 80s*. Columbus, OH: Charles E. Merrill.

Rogler, D. 2014. Assessment literacy: Building a base for better teaching and learning. *English Teaching Forum*, 52(3), pp. 2–13. US Department of State. Washington, DC: Bureau of Educational and Cultural Affairs.

Rowell, J.A. and Dawson, C.J. 1979. Cognitive conflict: Its nature and use in the teaching of science. *Research in Science Education*, 9(1), pp. 169–175.

Schoenfeld, A.H. 1987. What's all the fuss about metacognition? *Cognitive Science and Mathematics Education, 189,* p. 215.

Schunk, D.H. 2012. *Learning theories: An educational perspective* (6th ed.). Boston, MA: Pearson Education Inc.

Sebate, P.M. 2012. *The role of teacher understanding in aligning assessment with teaching and learning in Setswana home language* (Doctoral dissertation, University of South Africa, Pretoria).

Shakir, R. 2009. Soft skills at the Malaysian institutes of higher learning. *Asia Pacific Education Review, 10*(3), pp. 309–315.

Shapiro, S. and Leopold, L. 2012. A critical role for role-playing pedagogy. *TESL Canada Journal,* p. 120.

Shiah, Y.J., Huang, Y., Chang, F., Chang, C.F. and Yeh, L.C. 2013. School-based extracurricular activities, personality, self-concept, and college career development skills in Chinese society. *Educational Psychology, 33*(2), pp. 135–154.

Shuell, T.J. 1986. Cognitive conceptions of learning. *Review of Educational Research, 56,* pp. 411–436.

Smith, E. 2009. New models of working and learning: How young people are shaping their futures differently. *Research in Post-Compulsory Education, 14*(4), pp. 429–440.

Smith-Ruig, T. 2014. Exploring the links between mentoring and work-integrated learning. *Higher Education Research & Development, 33*(4), pp. 769–782.

Steenekamp, K. 2004. *The improvement of teaching practice in higher education* (Doctoral dissertation, University of Johannesburg, Johannesburg).

Stirling, A., Kerr, G., Banwell, J., MacPherson, E. and Heron, A. 2016. *A practical guide for work-integrated learning: Effective practices to enhance the educational quality of structured work experiences offered through colleges and universities.* Totonto, ON: Higher Education Quality Council of Ontario.

Telford, R.D., Cunningham, R.B., Fitzgerald, R., Olive, L.S., Prosser, L., Jiang, X. and Telford, R.M. 2012. Physical education, obesity, and academic achievement: A 2-year longitudinal investigation of Australian elementary school children. *American Journal of Public Health, 102*(2), pp. 368–374.

Van der Merwe, R. 2011. *Formative assessment in senior phase mathematics* (Doctoral dissertation, University of the Free State, Bloemfontein).

Watanabe, M. 2004. Networking, networking, networking, networking, networking. *Nature, 430*(7001), p. 812.

Wilson, N. 2009. *Impact of extracurricular activities on students.* Research paper, University of Wisconsin-Stout, Menomonie.

Wilson, S.M. and Peterson, P.L. 2006. *Theories of learning and teaching: What do they mean for educators?* Washington, DC: National Education Association.

Yorke, M. and Knight, P. 2006. *Embedding employability into the curriculum.* York: Higher Education Academy.

5 International context of higher education

Roles of higher education

Generally, the role of universities in improving graduate employability has earmarked it as a critical enabler for economic growth among both developing and developed countries worldwide (Yusuf *et al.*, 2009). Its role in research, development, innovation, and pedagogical approaches in developing graduates has rendered its function even more critical in our present day. Del Castillo (2000) describes higher education as a critical tool that proffers solutions to arising world problems by training citizens to be fit for service. This is because individuals' personal development and competencies are fostered on the basis of the quality of the educational processes they encounter during their university education (Mihail, 2006; Støren and Aamodt, 2010; Munap *et al.*, 2015). Consequently, individuals who benefit from these educational processes are fortified with the requisite skills, understanding, and competencies to handle significant positions in the construction industry in both design and supervisory capacities as well as function effectively in the society in which they find themselves. These educational processes also provide individuals (who in turn may become policy and lawmakers) with the much needed problem-solving abilities to effectively challenge arising issues in the work setting as well as their personal lives. Over the years, the relationship between higher education activities and employment has generated several debates which have led to the advent of several agreements, policies, debates, and declarations in certain parts of Europe and the world at large (Schomburg and Teichler, 2007). One of the major concerns raised was

> the increasing speed of turnover of knowledge required in jobs, the dramatic structural changes of the labour force in the wake of the introduction of new technologies and new managerial concepts, the globalization and Europeanization of the economy and society, the rapid 'massification' of higher education since about the mid-eighties in many industrialized societies, increasing unemployment etc.
>
> (Schomburg and Teichler, 2007)

These concerns necessitated the need for the urgent revamp and revision of higher education and its activities to produce adequately skilled graduates who are fit for purpose.

In addressing the needs of the future, the Accreditation Board for Engineering and Technology (ABET) insists that built environment graduates must be able to design a system effectively that can combat the arising environmental, socio-economic, ethical, and sustainability issues. They should also understand the ethical implications of their design and implementation choices as well as comprehend the possible impacts of engineering solutions in a specific context (economic, global, and environmental). Likewise, the Barcelona Declaration indicates present-day graduates should be able to understand the interaction between their work and immediate environment, adapt to current technology, work in multidisciplinary set-ups, and offer systemic and holistic approaches in solving arising problems. They should also be able to make meaningful contributions to technological and socio-economic policies, implement professional knowledge, and understand the demands of citizens with regard to the execution of technological and infrastructural services. Similarly, the United Kingdom Engineering Council asserts that graduates must act responsibly in a bid to improve their social lives and environment, be creative and innovative to maintain the quality of their society, and encourage stakeholder involvement to achieve societal growth.

The concept of globalization

The present-day technology brings into play the various dynamics associated with the information revolution that has influenced people's interaction with their environment. Over the years, there has been a significant increase in socio-economic relations among nations. This has prompted the elimination of local barriers as various nations strive to participate in the global movement, resulting in the saying, 'The world is gradually becoming one global village'. The consequences of this led to the advent of the term 'globalization'. Somehow, the developed nations of the world have been able to adjust to this societal change, mainly because of their distinct cultural, economic, historical, political, and sustainability features. Other nations, most especially the developing ones, are continually confronted by several challenges on how to strike a balance between buying into this global concept and promoting their cultural and economic values.

Consequently, the concept of globalization refers to a process of integrating local concepts onto a global scale by using innovative communication and information technology. Historically, it is the incorporation of local economies into the global economy via constant communication and the proliferation of technologies (Ngowi *et al.*, 2005). Globalization can also be seen as the constant amalgamation of several economies around the world via several indices and the continuous movement of both labour and technology (people and knowledge) through international borders. Cheng (2000) refers to globalization as the transmission, adaptation, and nurturing of technology, knowledge, behavioural types, societal values, and even cultures across nations of the world. Features peculiar to globalization include the global transfer of ideas and solutions, interflow of technological process, exchange of learning cultures, international partnerships and exchange, global networking, global use of international standards and benchmarks, and cultural interactions amongst others (Cheng, 2000). With the concept widely regarded as a speedy development in communication technology, several establishments and institutions now integrate the term 'global' in their vision and mission statement when drafting their operational policies and outlining their plans for nurturing future graduates. For others, globalization is also termed 'internationalization'. Hence, globalization deals with all spheres of life such as culture, economy, interaction, and most especially education (Scholte, 2008). In our present day, the concept of globalization is evident in the way the world currently operates in instances such as Internet usage for education and research purposes, international conferences, symposiums, visits and seminars, video conferencing across nations, web-based learning activities, and global exchange programmes (Jung and Rha, 2001).

As mentioned earlier, the globalization concept bodes well for individuals, organizations, and economic societies across different nations of the world as it offers numerous prospects of knowledge-sharing across various stages. Several advantages of globalization include promotion of international understanding, global exchange of intellectual assets, synergy among different countries, enhancing efficiency through global sharing, reception to cultural diversity across several nations and regions, mutual support across various cultural norms, encouragement of multi-way and multidimensional interaction, and communication threads (Cheng, 2000). Concurrently, the globalization concept has its downside in developing nations with regard to its negative impacts on their development. This has led to several social movements fighting against the globalization trends in these areas. Some disadvantages of globalization include a yawning and increased technological gap between developed and developing countries, the creation of more legitimate prospects of political and economic colonization of underdeveloped nations by more advanced ones, the exploitation of indigenous resources and cultures of less developed nations to benefit their more advanced counterparts, and a further increase of inequality and conflicts among cultures (Cheng, 2000).

Therefore, in our globally connected world (globalized world), there is the need for creative and responsive individuals who possess skills, knowledge, competencies, and values to handle key positions successfully. Considering the role of education in shaping society, the role of globalization places more pressure on educators and training providers. This increases the need for educators to place more emphasis on subject internationalization and re-conceptualization of their pedagogic methods in equipping learners with multidimensional knowledge, skills, and competencies to meet up with current trends. In order to achieve this, Cheng et al. (2004) offered several theoretical backgrounds to promote both the local knowledge and human development of learners. The theories of Cheng et al. (2004) suggest that the disparities peculiar to local communities (cultural and historical backgrounds) simply means that knowledge flow and processes vary from one local context to another. It is worth noting that five types of local knowledge exist in globalized education discussion. These are cultural knowledge, educational knowledge, economic and technical knowledge, human and social knowledge, and political knowledge (Cheng et al., 2004). Several theories are discussed in promoting local knowledge in globalization. They include the theory of amoeba, birdcage, crystal, DNA, fungus, and tree (Cheong, 2002).

Theory of amoeba

For countries or regions not so concerned with their cultural beliefs or heritage, the conservation of local values and assets may not be a priority. They may be primarily concerned with adjusting to the constantly evolving global environment and maintaining their financial stability in the face of international competition. Hence, the acceptance of this theory is necessary for these regions as it aims to utilize knowledge on a global scale with minimal issues. The theory of amoeba also posits that the process of global knowledge accumulation in a local setting is possible via improving local knowledge. However, whether this accumulated knowledge can endure is not a critical cause for concern.

   This theory opines that curriculum design should incorporate a holistic approach in accommodating global perspectives and knowledge so that education can be globalized to reap the benefits of global knowledge. Ultimately, the main aim of the theory is to ensure that individuals from the local community can think and act globally without any inhibitions or cultural constraints. Despite these strengths, the theory has certain limitations, including a potential tendency for individuals to lose their local identity and cultural values as during an overwhelming wave of globalization. On a broader scale, it is difficult to ascertain the outcomes of local contributions to ensure the development and integration of global knowledge (Cheong, 2002).

Theory of birdcage

One of the critical concerns of the globalization and modernization process is how to mitigate the overpowering global influences on local communities and societies. Hence, this theory is mostly considered as a key approach to combating the issue concerned with developing local knowledge. The theory of birdcage posits that developing local knowledge can work in tandem with global knowledge, although it calls for a limited form of interaction with the world at large to align with a certain thought process. This implies that the improvement and development of local ideas in a globalized educational sphere require indigenous ideas to check the quality of external knowledge to prevent negative global influences on the local society. Therefore, creating an educational curriculum that embraces indigenous emphasis even with the acceptance of global knowledge is of paramount importance in a globalized educational environment. Undoubtedly, loyalty to one's culture and norms should be a core aspect of education.

   The main aim of the theory is to develop local individuals who have a critical understanding of how the world works, yet still act and think indigenously with filtered global knowledge. This theory recognizes the place of local relevance in globalized education and helps to eradicate concerns of lost identity during modernization or international exposure. Principally, the designed indigenous framework may act as a buffer for locals during a case of overwhelming globalization. Despite these strengths, it could be challenging to build sociocultural boundaries to control the influence of global impacts whilst ensuring local relevance. This could be because of very close-knit boundaries that could prevent interaction with the external world or too loose boundaries that could lose their filtering and protection purposes (Cheong, 2002).

Theory of crystal

Some nations or regions may express concerns that complete exposure to globalization and modernization may lead to identity loss, prompting them to nurture indigenous knowledge in a particular aspect. Hence, this theory addresses this concern such that the fostering process requires the 'local ideas or seeds' to crystallize and embrace global knowledge. This theory posits that the primary design of the curriculum is aimed at identifying central-local values and traits and regarding them as the necessary seeds needed to accrue relevant educational resources and global knowledge. This is because in globalized education, having a grasp of what the local knowledge entails positively impacts the students' knack for accumulating knowledge on a global scale. One of the educational outcomes of this theory is to develop a local individual who can think and act locally while possessing some measure of global educational knowledge. In reality, it is possible to pinpoint the specific needs of the local community from the vast global knowledge and prevent over-globalization. One of the limitations of this theory is that it is difficult to identify good local seeds or norms that can be used in the crystallization and localization of global knowledge. Therefore, owing to one-way crystallization or localization, there is no clear contribution to global knowledge growth (Cheong, 2002).

Theory of deoxyribonucleic acid

When one explores the historical antecedents of Asian countries, several reforms have been adopted in eliminating flawed native customs and replacing them with innovative concepts obtained from outside the Asian continent, which this theory signifies. The theory of deoxyribonucleic acid (DNA) applies to this line of thinking. According to the DNA school of thought, prominence is placed on the adoption of core global information elements to substitute weakened local elements that can be gained in local innovations. This implies that improving and nurturing local knowledge entails largely replacing worthless local knowledge with key global knowledge through globalization, globalization education, or modernization.

In this case, the educational curriculum design is to be locally and globally specific to select the best elements from both. Therefore, it is necessary to have a critical understanding of both strong and weak elements in education. One of the educational outcomes of this theory is to develop an individual who possesses a rich blend of local and global elements and can function effectively in both the local and global context. This theory provides an avenue to learn and adopt insightful knowledge and elements as well as improve existing local developments without any local hindrances and burdens. One of its limitations is that it may be culturally difficult to identify strong and weak elements (Cheong, 2002).

Theory of fungus

Generally, owing to several factors such as geographical, socio-economic, and historical, the advancement of a region or nation may be dependent on some others. Hence, it is not a surprising trend that the development of local knowledge often relies on the global understanding of the nations on which they rely. This mode of fostering local knowledge best describes the theory of fungus. This theory posits that it is easier and more efficient to embrace pertinent aspects of global knowledge for individual and community development than to construct local ones out of nothing. In this theory, the educational training curriculum should aim to explain to students which aspect of global knowledge is beneficial to them as well as to their local community. One of the educational outcomes of this theory is to develop individuals who are adequately furnished with a variety of global knowledge types. In terms of strengths, adopting relevant elements of global knowledge is easier for smaller nations than building up local ones from scratch. However, a limitation of this theory is the over-dependence on external knowledge because of its lack of clear local identity and vision for growth (Cheong, 2002).

Theory of tree

Several Asian nations with established cultural heritages, beliefs, and traditions are making extra efforts to maintain their local identities and cultural norms as they face the challenges of globalization. This theory posits that local knowledge should be reinforced in local traditions and must embrace valuable resources globally to experience local knowledge growth, both inwards and outwards. One of the educational outcomes of this theory is to develop individuals who are internationally driven and can act and think locally while developing globally. One of the strengths of this theory is the ability of local communities to improve their cultural identity by interacting with external resources, thereby leading to local development. Nevertheless, if local community roots are deficient, individuals, as well as their local community, will struggle to experience exponential growth.

The theories discussed previously all have their characteristics and peculiar strengths and weaknesses and these may vary according to various contexts (nations, societies, circumstances, and perspectives). Hence, it is difficult to conclude that anyone is superior to another. One thing is certain: these theories can offer researchers, policymakers, and training providers alike the opportunity to conceptualize and implement strategies of maximizing the advantages of globalization as well as fostering local knowledge. The theories of birdcage, crystal, and tree have a weaker global dependence, but a stronger local orientation. By contrast, both the amoeba and fungus theories exhibit a greater global dependence, but have the least local orientation. However, the theory of DNA relies on two groups (Cheong, 2002).

A comparative analysis of employability skills in selected countries

The United Kingdom

The United Kingdom is an amalgamation of four nations: England, Northern Ireland, Scotland, and Wales. After the Second World War, the economy of this region experienced a significant period of stagnation, whereas its neighbours experienced significant growth. Owing to the advent of the first industrial revolution during the 19th century, its economy was predominantly based on industries and manufacturing hubs till the 1980s. However, since the 1990s, the economy has centred more on the servicing industries that prompted a variety of different skills sets from the workforce (Keating *et al.*, 2002). Moreover, non-technical skills in this region were previously regarded as core skills. However, the term 'key skills' (relevant abilities to an individual's life and career) is currently adopted. The integration of other skills and competencies has seen employers refer to them as employability skills (NCVER, 2003). Over time, the importance of improving these employability skills have been underscored by government institutions such as the Department of Trade and Industry (DTI) and the Department for Education and Skills (DfES) as well as employer organizations and establishments. The partnership between the DTI and DfES paved way for the then Secretary of State for Education and Employment, David Blunkett, to institute the 'Key Skills National Qualifications' in 1997, which highlighted numeracy, information technology application, and effective communication as critical competencies for graduate success (Turner, 2002). Apart from these three basic proficiencies, the British Government advocated for three wider key skills to be regarded as crucial to enable individuals to achieve self and societal satisfaction. These skills are problem-solving, lifelong learning and teamwork abilities (Turner, 2001). Furthermore, Turner confirmed that the British government formulated several policy documents (such as the Lifelong-learning Green Paper) which aimed at encouraging the development of skills among young individuals and adults to remain employable.

With reference to the educational system in the United Kingdom, there is a continuous effort to make its economy as competitive as that of other member states of the Organisation for Economic Co-operation and Development (OECD). One of the critical ways of achieving this was to raise the skill levels of its citizenry to ensure an adaptable, industrious, creative, and productive workforce. In achieving this, the Department of Education and Skills is a sector tasked with the enactment of lifelong learning to ensure an all-inclusive society as well as a competitive economy (Department of Education and Skills, 2005). Also, the UK education system is divided into primary education, secondary education, and further education. On completion of primary education, students attend five years of secondary education and training before sitting for a General Certificate of Secondary Education (GCSE), which paves the way for an option to undertake further education at colleges (specialist schools and apprenticeships schemes).

The United States

Like the United Kingdom, there has been an evolution in the United States from manufacturing to being service-oriented and performance-driven. This has increased the need for a new set of skills and competencies from prospective employees (Barker, 2000). In the long run, this paradigm shift benefits those individuals who possess the relevant knowledge, education, and abilities that meet the increasing and ever-changing demands of the world of work. Over time, several Statutes and Acts have been instituted and implemented in the United States to advocate, implement, and oversee all levels of education (vocational and technical training) so that graduates can respond positively to the needs of employers upon graduation (Silverberg *et al.*, 2004). The advent of these bills has further strengthened the academic content and activities of higher education to ensure students' readiness for the industry as recommended by the Secretary's Commission on Achieving Necessary Skills (SCANS). The SCANS reports emphasize the role and significance of higher education activities in developing SCANS skills as well as the role of employers in inculcating these skills. Educators and their activities are hence critical in providing individuals with the ability to understand the relationship between lecture-room experience and real-life settings. In the United States, there is a well-structured public schooling set-up in place with the majority of higher education centres privately owned and catered for. Since the 1990s, there have been several calls and policies to reform the public secondary schools so that they align themselves with the dynamism of the workplace to produce individuals who are fit for purpose. In achieving this reform process, several of the strategies adopted were as follows: establishing technical programmes to bolster the technical careers of students; proper planning of curricula to allow students ample time to participate in active learning (field trips and laboratory activities); organized career activities to prepare students for employment opportunities; and skills assessments platforms to successfully monitor the competency level of students before graduation, amongst others (Keating *et al.*, 2002). After 12 years, students who finish high school (complete year 12) are awarded a high school diploma certificate. With this certificate, learners are eligible to enrol at any college, technical school, or university for further education.

Over time, several groups and bodies have conducted various research studies on employability skills required for industry success. For example, the American Society for Training and Development (ASTD) commissioned an employability baseline study which enumerated six employability skills of high importance for graduates to succeed. They include basic numeracy skills, basic communication skills, teamwork skills, leadership skills, development skills, and adaptability traits (Overtoom, 2000). This study by the ASTD was further encouraged and funded by the US Department of Labour, leading to the establishment of SCANS in

1990. Its formation was based on the need to identify the various key skills and competencies needed by school leavers to thrive in the workplace. Stakeholders from the education sector, labour market, and even the government formed part of the commission tasked with the purpose of establishing a set of generic competencies needed for societal growth (Richens, 1999). In 1992, SCANS developed a certain set of essential skills in developing a competitive workforce. These included personal qualities (integrity, positivity, self-esteem, and confidence), thinking skills (problem-solving, creativity and high reasoning), and basic skills (communicative skills). Likewise, Packer and Brainard (2003) identified key competencies that are pivotal for employee success. They are information skills, technological abilities, planning skills, adaptability, and interpersonal proficiencies (Packer and Brainard, 2003).

According to Gbomita (1997), the SCANS skills increasingly gained prominence in the United States, positively influencing the workplace and work-readiness training schemes such as the School-to-Work Opportunities Act, which was formed to ease students' transition from lecture room to work setting (Gbomita, 1997). A survey to assess the strength of SCANS skills was carried out in 1998 by the School-to-Careers Professional Development Centre (STCPDC) in Las Vegas, Nevada. Researchers received more than 400 responses to a questionnaire with the majority of the respondents endorsing the validity of the SCANS list. Notably, interpersonal skills and resourcefulness were highlighted by employees (Richens, 1999).

Australia

For a long period, the Australian economy has enjoyed a somewhat reasonable spell of stability which has resulted in low inflation, a high growth index, an efficient government system, and a very flexible and competitive labour market. With a very strong information communication technology (ICT) base as well as diverse functional developmental policies and innovations, the nation boasts one of the fastest-growing economies in the world. Like the United Kingdom and the United States, the nation's innovativeness and labour market changes have necessitated the need for reforms and readjustments to the nation's training sector to respond to new skills' demands. These reforms advocated for employer involvement in the structuring and dissemination of training to create a holistic level of skills development. It is worth noting that the Australian national training systems are particularly consistent, industry-driven, responsive, competency-based, and quality assured to meet the dynamic needs of employers.

Regarding the nation's educational system, students are awarded an upper secondary certificate upon completion of their compulsory secondary schooling. Afterwards, they are presented with the option to proceed to higher education, vocational education training (VET), or vocation education centres to commence an apprenticeship scheme. Mainly controlled and directed by the government, the largest providers of VET programmes in Australia are the Technical and Further Education (TAFE) institutions. These training providers and organizations help to offer industry-specific training schemes that can yield accredited vocational qualifications to individuals who undertake them (Misko, 2006). A standout feature of the nation's educational system is the competency-based approach that pays more attention to the learners' outcome rather than the efforts and resources invested in the training system. These learning outcomes act as a feedback mechanism in checking whether individuals have attained the expected competency levels. There are four strategies adopted in assessing the level of employability skills in the Australian educational space: workplace valuation, portfolio assessment, holistic findings, and standardized instrumental assessment (Curtis, 2004).

Discussions on employability skills development began in Australia during the 1980s and were revisited in the early 1990s owing to the dynamics of evolving industrial initiatives. Like the United Kingdom and the United States, the nation formulated Acts and Reforms to develop employability skills. Consequently, the Australian Education Council Mayer Committee adequately addressed these skills need by instituting the Mayer key competencies. This set includes problem-solving, information and communication technology (ICT) proficiencies, numeracy skills, teamwork abilities, communication skills, planning and organizing skills, and analytical reasoning. These employability skills can be developed via formal and informal approaches such as homes, job-settings, educational set-ups, social gatherings, and leisure activities.

The Report also confirms that employability skills can be promoted through educational training systems via three main strategies. They include integrating employability skills as part of the learning programme; developing employability skills to meet vocational and technical requirements; and developing employability skills using informal portfolios. In further examining the various skills and competencies required for work success, Crowley *et al.* (2000) carried out their studies in the building and construction industry in New South Wales through the help of interviews. The study was conducted with a cross section of employees requested to describe their daily work schedule. The following skills and competencies were indicated as essential for job success: communicative skills, teamwork abilities, literacy, and organizational abilities (Crowley *et al.*, 2000). In 2006, the Allen Consulting Group also commissioned an employability skills study based on 350 companies which belonged to a variety of sectors, including the construction, manufacturing, and the IT industry. From the interviews conducted among employees from the respective companies, it was revealed that to remain employable and competitive, employees need to possess the following: good interpersonal skills and values, literacy, numeracy skills, communication skills, teamwork abilities, and a willingness to learn. Like the Allen Consulting Group, the NCVER conducted two different baseline studies. The first, conducted by Dawe (2002), surveyed various approaches to incorporate generic skills and competencies into VET programmes. Ten industry sectors were interviewed: administration, agriculture, community services, construction, engineering, entertainment, hospitality, IT, retail, and the Australian Red Cross Blood Service. Three competencies and skills were found to be significant across the interviewed sectors: communication skills, teamwork abilities, and adequate knowledge of their workspace (health, safety, and security measures). Other competencies viewed as important were work ethics and professionalism as well as interpersonal skills (Dawe, 2002).

The second baseline study conducted by Clayton *et al.* (2003) surveyed six training firms: Campaspe College of Adult Education, Centrelink Agent, the Spencer Institute of TAFE, the University of Melbourne's Burnley College, and two senior secondary colleges from the Australian Capital Territory. From the results, it was discovered that the Mayer key competencies were acknowledged in the various organizations. However, additional skills such as adaptability and personal values were identified as vital for workplace success (Clayton *et al.*, 2003). Likewise, the Australian Chamber of Commerce and Industry and the Business Council of Australia commissioned a baseline study to identify key skills required for job success. Apart from reviewing extant literature from Australia and abroad, the group interviewed stakeholders from several sectors. Their findings went beyond the Mayer key competencies as they integrated both personal attributes and general employability skills. The following eight skills were found in the ACCI/BCA framework: communication skills, initiative and enterprise, lifelong learning, planning and organizational skills, problem-solving skills, self-management skills, teamwork abilities, and technological competencies. This framework also factored in several personal attributes for workplace success, including adaptability, honesty, reliability, and loyalty (NCVER, 2003).

New Zealand

In New Zealand, studies on employability skills and competencies commenced in 1984, following a review of the Core Curriculum for Schools. This heralded the publishing of several critical reports such as the Curriculum Review report (1985–1987), the Learning and Achieving report (1986), and the Department of Education's draft National Curriculum statement (1988). According to Kelly (2001), these reports were designed to provide a robust curriculum to enable learners to develop the requisite skills that would help improve them as well as their immediate society. In 1990, the nation's Education Amendment Act was passed which followed the formation of the New Zealand Curriculum Framework (NZCF) in 1993. The NZCF framework, which applies to the curriculum of New Zealand, provides eight sets of skills deemed essential to enable students to achieve their set out employment goals (Kelly, 2001). These skills include communication skills, information management and technology, numerate skills, physical skills, problem-solving abilities, self-management skills, social skills, and work ethics. A few years later, the NZCF decided to make modifications based on the recommendations of a 'curriculum stocktake', which was commissioned by the nation's Ministry of Education to review the extent to which the set of skills were being measured by the nation's educational system. Founded in 2003, the Curriculum/Marautanga project was instituted to make necessary adjustments to the national curriculum and to offer educators the necessary expertise in fostering the skills and competencies outlined earlier. In attaining the aims of the Curriculum/Marautanga scheme, several key stakeholders, including students and educators, were involved in employability skills discussions. Concurrently, the Ministry of Education established a White Paper containing various conceptual frameworks and explaining the various competencies employees need to possess to succeed across different fields. One of the competency frameworks was the OECD's Definition and Selection of Competencies (DeSeCo). Hence, this project was adapted into the context of New Zealand (Rutherford, 2005). In 2005, educators and policymakers received a draft version of the key competency framework, which resulted in five different groups of competencies and skills. They include adequate knowledge of chosen fields, a holistic understanding of dynamic ideas, interpersonal skills, organizational abilities, and critical thinking skills (Young and Chapman, 2010). Similarly, findings from the studies conducted by the New Zealand Employment Service found that the majority of the skills and competencies deemed essential by employers were interpersonal in nature. They include communication skills, IT skills, cooperative (teamwork) skills, critical thinking, and creativity skills (Kelly, 2001).

Canada

Generic skills and competencies at the workplace gained prominence in 1974. However, the need for an agreed framework required for industry or workplace success only gained ground a few years later in the late 1980s and early 1990s. Like other developed countries, the Canadian government commissioned many studies to identify the competencies for workplace success. Consequently, the Conference Board of Canada's Corporate Council conducted a skill identification project, which began with an extant review of literature to ascertain employer expectations with regard to skills and competencies. This first stage of research yielded a draft set of employability skills, according to McLaughlin (1992). The second stage required the validation of the draft via the contributions of stakeholders (educators, employers, society groups, and government officials) who were mandated to peruse the framework and comment where necessary. Findings from the report established that employers value the following competencies: communication skills, lifelong learning, IT skills, personal qualities, critical thinking skills, adaptability, and creativity (McLaughlin, 1992). In ensuring the development of these traits and attitudes sought after by employers and recruiters, the Employability Skills Profile (ESP) was established. The ESP was classified into three groups which were considered equally important. These groups were academic skills (lifelong learning, critical thinking, organizing skills, and communication skills), personal skills (honesty, adaptability, integrity, confidence, and ethical responsibility), and teamwork skills (people skills and leadership skills).

Years later, the ESP was updated through the Employability Skills 2000+ scheme, which factored in some additional skills taking into consideration the dynamic changes which had occurred because of certain global economic factors. These additional skills include fundamental skills (information management and problem-solving skills), personal management skills (adaptability and lifelong learning principles), teamwork abilities (solving problems as a team leader and a member of a team), and ethical principles (integrity, ethical considerations, and taking responsibility) (Young and Chapman, 2010).

Other European countries

After the Second World War (1939–1945), several European nations saw the need for skill development and economic growth as they rebuilt their depleted economies. As time went on, advancement in science and technology, coupled with competitive markets, required learners to possess a certain level of skills and competencies required for economic success. Owing to the diversity and vastness of the European continent, it was not until the late 1990s before reports on employability skills and competencies gained significant prominence. However, in 2001, the European Commission commissioned a study to examine the various definitions of generic competencies (key competencies) according to its member nations. Subsequently, questionnaires were designed and distributed to member nations (Austria, Belgium, Denmark, France, Finland, Germany, Greece, Italy, Luxembourg, the Netherlands, Northern Ireland, Portugal, Spain, Sweden, Northern Ireland, and the United Kingdom), through the Eurydice European Unit. The study aimed to examine the necessary skills and competencies needed for workplace success. Findings from the study revealed that the skills and competencies varied across the member nations. However, the United Kingdom (with the exclusion of Ireland) was the only respondent to make a distinction between generic competencies and subject-specific competencies in its academic curriculum. Notably, all member nations, whether directly or indirectly, encouraged creative and critical thinking, promoted academic skills, and emphasized teamwork skills as well as communicative abilities (Eurydice, 2002; Gibbons-Wood and Lange, 2000).

Concurrently, the European Council converged in Lisbon, Portugal, to deliberate on several ways and approaches by which the European community could become a global leader with regard to knowledgeability and employability. This led to the formation of a group tasked with recognizing the various skills and competencies that would hand Europe the initiative in achieving its outlined goals. In boosting Europe's economy, eight generic competencies were identified for its learners to possess upon graduation: communication in native languages, proficiency in foreign languages, numeracy abilities, ICT skills, lifelong learning, interpersonal skills and responsibilities, entrepreneurial abilities, and cultural awareness (Eurydice, 2002; Young and Chapman, 2010).

As seen from the foregoing text, the employability skills discussion is a well-deliberated topic because present-day employers seek graduates with relevant skills, traits, and competencies to meet industry needs after graduation. It has also become one of the most researched areas and this is evident in the volume of researchers and nations who have engaged with the subject of employability skills. Table 5.1 presents an extensive list of various studies that have contributed to the employability skills discussion globally in recent years.

Table 5.1 Summary of previous researches on employability skills

No.	Author(s) and year	Country	Purpose of the study
1	Husain *et al.* (2010)	Malaysia	Key employability skills were determined from the perspective of industry employers
2	Mitchell *et al.* (2010)	The United States	Business educators were required to identify the various soft skills needed for employment success
3	Davies *et al.* (2011)	The United States	Key skills required to function effectively in the next decade were determined from the perspective of industry professionals
4	Johari *et al.* (2011)	Malaysia	Critical skills required to thrive in the industry from the perspective of industry employers
5	Lim *et al.* (2011)	Malaysia	Competencies required for job success from the perspective of industry employers
6	Warraich and Ameen (2011)	Pakistan	Skills required for job success from the perspective of both graduates and industry employers
7	Blom and Saeki (2011)	India	Key skills required for job success from the perspective of industry employers
8	Messum *et al.* (2011)	Australia	The top ten employability skills were determined from the perspective of employers
9	Yuzainee *et al.* (2011)	Malaysia	Engineering employers were required to identify critical employability skills
10	Robles (2012)	The United States	Business employers' identified their top ten soft skills
11	Nickson *et al.* (2012)	The United Kingdom	Key skills required for job success from the perspective of fashion retailers
12	Omar *et al.* (2012)	Malaysia	To determine the employability level of graduates from the perspective of advertising company employers
13	Poon (2012)	The United Kingdom	To determine the employability skills of real estate graduates from the perspective of human resource managers
14	Srivastava and Khare (2012)	India	To determine the employability of recruits from the perspective of human resource managers
15	Selvadurai *et al.* (2012)	Malaysia	Key skills required for job success from the perspective of social employers

(Continued)

Table 5.1 (Continued)

No.	Author(s) and year	Country	Purpose of the study
16	Jackson (2013)	Australia	Key skills required for job success from the perspective of business students
17	Griffin and Annulis (2013)	The United States	Key skills required for job success from the perspective of students and manufacturing employers
18	Seth and Seth (2013)	India	Top executives and employers determined critical employability skills
19	Balcar *et al.* (2014)	Czech Republic	Key skills required for job success from the perspective of the Czech labour force
20	Saad and Majid (2014)	Malaysia	Skills required for job success from the perspective of ICT employers
21	Chavan and Surve (2014)	India	Key employability skills required for job success from the perspective of employers across different sectors
22	Ajwad *et al.* (2014)	Uzbekistan	To determine the skill-gap and possible employability skills required for job success from the perspective of employers across different sectors
23	Mirza *et al.* (2014)	Pakistan	Key skills required for job success from the perspective of industrial sector employers
24	Ho (2015)	Taiwan	Key competencies required for job success from the perspective of employers
25	Furnell and Scott (2015)	The United Kingdom	Key skills required for job success from the perspective of freshly employed bioscience graduates
26	Pradhan (2015)	India	Key employability skills required for job success
27	Messum *et al.* (2015)	Australia	Key employability skills required for job success from the perspective of New South Wales health managers
28	Ramadi *et al.* (2015)	The Middle East and North Africa (MENA) region	Examining industry expectations of graduate skills from industry professionals
29	Yang *et al.* (2015)	China	Key competencies required for job success from the perspective of graduates themselves
30	Lim *et al.* (2016)	Malaysia	Key personal qualities and employability skills required for job success from the perspective of accounting employers

31	Mansour and Dean (2016)	Morocco, Europe, and the United States	Key personal qualities and employability skills required for job success from the perspective of employers and university stakeholders
32	Vashisht *et al.* (2016)	India	Key abilities required to achieve technical functions in the world of work
33	Rahmat *et al.* (2016)	Malaysia	To determine the various dimensions of employability skills required for job success from the perspective of employers from the electrical and electronics industry
34	Qomariyah *et al.* (2016)	Indonesia	To determine the essential employability skills required for job success from the perspective of graduates and employers
35	Salleh *et al.* (2016)	Malaysia	To emphasize the importance of employability skills among graduate architects from the perspective of industry employers
36	Messum *et al.* (2016a)	Australia	To review the importance of employability skills among graduates
37	Messum *et al.* (2016b)	Australia	Key employability skills required for job success from the perspective of health science graduates
38	Rajapakse (2017)	Sri Lanka	Key competencies required for job success from the perspective of employers
39	Ghazali and Bennett (2017)	Malaysia	Key competencies required for job success from the perspective of employers
40	Singh *et al.* (2017)	India	Key competencies required for job success from the perspective of employers
41	Ayoubi *et al.* (2017)	Syria	Key competencies required for job success from the perspective of employers, policymakers, and higher education stakeholders
42	Clarke (2018)	Australia and the United Kingdom	Explore and explain the concept of graduate employability

Review of employability skills

The present-day dynamics of the construction industry require built environment graduates to be adequately skilled and industry ready. These graduates are also regarded as the future custodians of the industry as they are considered as vehicles for the integration of techniques and technology to deal with arising industry issues and challenges (Farooqui and Ahmed, 2009; Tatum, 2010). With an increase in technological advancement and its prevalence in industry activities, the importance of employability skills cannot be overemphasized. Apart from possessing strong academic skills, graduates are also required to possess non-academic skills. A review of key employability is conducted next.

Communication skills

This skill refers to the fluency of graduates in speaking, reading, listening, and writing effectively. It is worth noting that the construction industry is one that is built on relating to professionals and team members, amongst others. Therefore, graduates who lack proficiency in communicating effectively may be disadvantaged in job execution and professional development (Gardner *et al.*, 2005; Farooqui and Ahmed, 2009; Ariana, 2010). Communication skills can be developed among students through writing and presenting reports, working in focus groups, and role-play, amongst others.

Teamwork skills

This is the ability and willingness to work and cooperate with people or personalities from various backgrounds (social and cultural) in achieving a common goal to maintain productivity (Washer, 2007; Samavedham and Ragupathi, 2012). Apart from developing their communication skills, graduates with teamwork skills can effectively benefit the industry by displaying varying levels of ingenuity, innovativeness, and quality decision-making. Teamwork skills can be developed among students through group discussions, learning sets, and group projects, amongst others.

Problem-solving skills

This is a fundamental skill that enables graduates to exhibit creativity and practicality in proffering solutions to arising industry problems (Wickramasinghe and Perera, 2010; Finch *et al.*, 2013; Ahn *et al.*, 2012; Reid and Anderson, 2012; Jackson and Chapman, 2012). It also involves recognizing the problem, picking a viable solution, and implementing the solution discovered. Graduates with problem-solving skills take responsibility for ensuring the actualization of projects by ensuring a better strategy in solving problems. In preparing built environment students to be effective problem solvers, both higher education and educators need to understand the need for this skill to be developed. Employers seek graduates who are problem-solvers as they are regarded as more productive. There are various processes towards problem-solving; however, the one put forward by Pokras (1995) stands out. He advocates six steps which include identifying the problem, categorizing the problem, analysing the reason for the problem, exploring possible solutions to solve the problem, making a choice to resolve the problem, and implementing an action plan to solve the problem (Pokras, 1995).

Self-management skills

These sets of skills refer to the capacity to manage one's time and self responsibly in achieving organizational goals. As a component of emotional intelligence, self-management provides individuals with knowledge about their possible strengths and weaknesses (Steyn and Van Staden, 2018). It is also described as the ability to keep negative emotions in check. Individuals who possess this skill are optimistic, proactive, calm, and positive despite setbacks: these are key traits in achieving industry success (Valentine *et al.*, 2014). Certainly, individuals who possess this skill can create a trustworthy and fair environment. Other components of self-management include self-awareness, goal setting, personal drive, resilience, time management, balancing work/life issues, creativity, and self-confidence (Steyn and Van Staden, 2018).

Adaptability and risk-taking skills

Employers also hold in high esteem those graduates who are open to new ideas and adapt easily to new surroundings (Love *et al.*, 2002; Rawlins and Marasini, 2011; Ahn *et al.*, 2012). Risk-taking also involves identifying optional ways of achieving objectives while recognizing the possibility of failure and monitoring progress towards a predetermined set of objectives (Evers *et al.*, 1998). Furthermore, to achieve a constructive result, a graduate must be able to take risks and be willing to deal with the resulting consequences of risk-taking (DuBrin, 2007).

Critical thinking/creative thinking skills

Graduates who possess this skill are regarded as creative and critical thinkers because creativity results in change and innovation. This set of skills provides graduates with the ability to apply well-thought ideas to initiate change and provide novel solutions to arising industry problems (Kilgour and Koslow, 2009; Samavedham and Ragupathi, 2012; Reid and Anderson, 2012; Jackson and Chapman, 2012). According to DuBrin (2007), there are five stages in the creative thought process: problem recognition (which implies the problem is worth solving); immersion (which implies concentration on the identified problem); incubation (which implies the subconscious meditation on the problem); insight (which implies envisioning of a possible solution), and verification and application (which implies the assembling of possible materials to solve the problem) (DuBrin, 2007).

Organizing and time-management skills

This skill is critical for graduates as it provides the ability to set priorities, think critically, and plan around those priorities. This set of skills also involves multitasking as well as managing time effectively to meet deadlines promptly (Rawlins and Marasini, 2011; Jackson and Chapman, 2012). Fritz *et al.* (2005) provide the following ten significant ways of managing time effectively: drawing up an achievable plan, identifying priorities, setting of realistic targets, adequate planning, establishing deadlines, delegating tasks when feasible, planning meetings carefully and timeously, developing procedures to accumulate data, grouping similar work tasks, and scheduling some personal time daily (Fritz *et al.*, 2005).

Lifelong learning skills

This set of skills provides graduates with the ability to take responsibility for their learning process and solicit feedback from their supervisors, peers, and even mentors (Casner-Lotto and Barrington, 2006). Graduates with these skills understand that learning can lead to skills, competencies, values, knowledge, attitudes, flexibility, empowerment, self-awareness, and even to self-confidence (Honey, 2001). Hence, the possession of this skill can lead to the acquisition of other fundamental and job-related skills. Lifelong learning can occur through formal education, informal discussions, basic education, adult education, symposiums, open-day forums, in-service training, and seminars (Casner-Lotto and Barrington, 2006).

Individual skills

These sets of skills and strengths are significant for all graduates to possess as they can improve their productivity in the workplace. They are regarded as traits that aid individuals in dealing with situations they encounter as well as arising problems. These traits include maintaining a high energy level, maintaining a good rapport with other professionals, managing time, motivating oneself to function optimally, exhibiting commitment, working independently, responding positively to criticism, working under pressure, being honest, and showing enthusiasm (Lievens and Sackett, 2012; Ayarkwa *et al.*, 2012; Finch *et al.*, 2013; O'Leary, 2017).

Civic-mindedness

These sets of skills provide graduates with the sense that their education will prepare them to influence their communities positively (Evers *et al.*, 1998). Being civic-minded implies an understanding of education and its implications to serve as a citizen. Shiarella *et al.* (2000) also argue that community service experiences are pivotal to the holistic education of students. It is therefore one of the primary missions of universities to educate students to enable them to meet the needs of the construction industry as well as their immediate communities (Colby *et al.*, 2003). Through work-integrated learning, problem-based learning, experiential education, and service-learning opportunities, students can understand their communities and can develop several skills and values, including communication skills, teamwork skills, leadership abilities, organization skills, conflict resolution, time management skills, and personal values (Evers *et al.*, 1998). Along with these skills, students can also be exposed to the concept of corporate social responsibility (CSR) which implies 'giving back' as this is critical to their success as employees as well as corporate (socially responsible) citizens (Colby *et al.*, 2003). Therefore, students need to be engaged in learning prospects to develop civic will, as this will improve their self-worth in their place of work and communities alike.

Technology skills

These sets of skills provide graduates with the ability to utilize appropriate technological tools to solve arising industry problems. With the industry increasingly driven by technological advancement and innovations, graduates are required to familiarize themselves with up-to-date software and applications in order to be relevant (Russell *et al.*, 2007; Tatum, 2010; Arain, 2010; Ahn *et al.*, 2012).

Numeracy skills

These sets of skills provide graduates with the ability to manipulate number functions and data with relative ease. Number functions include estimations, costing, measuring, calculations, and numeric problem-solving (Durrani and Tariq, 2012). Numeracy skills also refer to the capacity to utilize computer software (spreadsheets) in carrying out calculation tasks and solving arithmetic problems (Jackson and Chapman, 2012).

Entrepreneurship skills

These sets of skills provide the capacity to show ambition by venturing into work-related opportunities and businesses despite the risk involved. It also involves the ability to plan, build, and explore business activities that can eventually lead to self-employment. Good entrepreneurs display a willingness to listen and learn, creativity, perseverance, confidence, and risk-taking (Nowiński and Haddoud, 2019).

Leadership skills

These sets of skills provide the ability to take charge or lead a team during work activities. According to Graham *et al.* (2009), it is necessary for graduates to display certain elements and dimensions of leadership traits such as exhibiting confidence, displaying knowledge, and problem-solving when handed the opportunity to oversee a given task. It also involves taking responsibility for team members' knowledge base and well-being and ensuring that work goals are met (Conrad and Newberry, 2012).

Ethical skills

These sets of skills provide the ability for graduates to uphold organizational statutes and policies as well as maintain set-out building codes and regulations (Jackson and Chapman, 2012). Ethical skills also involve the exhibition of key moral principles when handling work activities. These moral principles include accepting responsibility, honesty, dignity, and fairness (Mat and Zabidi, 2010).

Individual values

These include significant features and traits all graduates must possess, whether adequately skilled or not, as they are essential to enhancing workplace productivity. Individual values, also known as soft skills, help to promote excellent working conditions and harmony. Such values include self-confidence, hard work, attention to detail, creativity, time-management, enthusiasm, positive outlook, honesty, professionalism, social sensitivity, commitment, and reliability (Lievens and Sackett, 2012; Finch *et al.*, 2013). These employability skills have also been regarded as essential by several engineering accreditation bodies across the world. Owing to the advent of the 4IR, most of these skills are no longer sufficient as they need to be enhanced, as discussed extensively in Chapter 3.

Summary

This chapter reviewed the various employability and generic competency frameworks from an international context. The chapter also examined the role of graduates in socio-economic development according to several international legislations and policies. Additionally, the concept of globalization and its theories were accentuated. Furthermore, a comparative employability skills analysis was conducted in selected countries. This chapter also revised several international research studies conducted on employability and reviewed various employability skills required for graduate success. The next chapter will provide an overview of the educational system in South Africa as well as various legislations and policies that have some level of influence on the education and training process.

References

Ahn, Y.H., Annie, R.P. and Kwon, H. 2012. Key competencies for US construction graduates: Industry perspective. *Journal of Professional Issues in Engineering Education and Practice*, 138(2), pp. 123–130. doi: 10.1061/(ASCE)EI.1943-55410000089.

Ajwad, M.I., Abduloev, I., Audy, R., Hut, S., Laat, J.D., Kheyfets, I., Larrison, J., Nikoloski, Z. and Torracchi, F. 2014. *The skills road: Skills for employability in Uzbekistan.* Washington, DC: World Bank.

Arain, F.M. 2010. Identifying competencies for baccalaureate level construction education: Enhancing employability of young professionals in the construction industry. *Proceedings of the Construction Research Congress 2010: Innovation for Reshaping Construction Practice*, Construction Research Congress 2010, May 8–10, 2010, American Society of Civil Engineers (ASCE), Banff, AB.

Ariana, S.M. 2010. Some thoughts on writing skills. *Annals of the University of Oradea, Economic Science Series*, 19(1), pp. 134–140.

Ayarkwa, J., Adinyira, E. and Osei-Asibey, D. 2012. Industrial training of construction students: Perceptions of training organizations in Ghana. *Education Training*, 54(2/3), pp. 234–249. http://dx.doi.org/10.1108/00400911211210323.

Ayoubi, R.M., Alzarif, K. and Khalifa, B. 2017. The employability skills of business graduates in Syria: Do policymakers and employers speak the same language? *Education + Training*, 59(1), pp. 61–75. https://doi.org/10.1108/ET-10-2015-0094.

Balcar, J., Janickova, L. and Filipová, L. 2014. What general competencies are required from the Czech labour force? *Prague Economic Papers*, 2014(2), pp. 250–265.

Barker, J. 2000. School counsellors' perceptions of required workplace skills and career development competencies. *Professional School Counselling*, 4(2), pp. 134–139.

Blom, A. and Saeki, H. 2011. Employability and skill sets of newly graduated engineers in India. *Policy Research Working Paper Series*. The World Bank South Asia Region.

Casner-Lotto, J. and Barrington, L. 2006. *Are they really ready to work? Employers' perspectives on the basic knowledge and applied skills of new entrants to the 21st century US workforce*. Washington, DC: Partnership for 21st Century Skills.

Chavan, R.R. and Surve, A.Y. 2014. Assessing parameters of employability skills: An employers' perspective. *Asian Journal of Management Research*, 5(2), pp. 254–260.

Cheng, B.S., Chou, L.F., Wu, T.Y., Huang, M.P. and Farh, J.L. 2004. Paternalistic leadership and subordinate responses: Establishing a leadership model in Chinese organizations. *Asian Journal of Social Psychology*, 7(1), pp. 89–117.

Cheng, Y.C. 2000. A CMI-triplization paradigm for reforming education in the New Millennium. *International Journal of Educational Management*, 14(4), pp. 156–174.

Cheong, C.Y. 2002. *Fostering local knowledge and wisdom in globalized education*. Multiple Theories Centre for Research and International Collaboration. Hong Kong: Hong Kong Institute of Education.

Clarke, M. 2018. Rethinking graduate employability: The role of capital, individual attributes and context. *Studies in Higher Education*, 43(11), pp. 1923–1937.

Clayton, B., Blom, K., Meyers, D. and Bateman, A. 2003. *Assessing and certifying generic skills: What is happening in vocational education and training?* Adelaide, Australia: The National Council for Vocational Education Research.

Colby, A., Ehrlich, T., Beaumont, E. and Stephens, J. 2003. *Educating citizens: Preparing America's undergraduates for lives of moral and civic responsibility*. San Francisco, CA: Jossey-Bass.

Conrad, D. and Newberry, R. 2012. Identification and instruction of important business communication skills for graduate business education. *Journal of Education for Business*, 87(2), pp. 112–120.

Crowley, S., Garrick, J. and Hager, P. 2000. Constructing work: Generic competencies and workplace reform in the Australian construction industry. *Future Research, Research Futures: Proceedings of the Third National Conference of Australian Vocational Education and Training Research Association (AVETRA)*, Sydney, Australia.

Curtis, D. 2004. The assessment of generic skills. In J. Gibb (Ed.), *Generic skills in vocational education and training: Research readings* (pp. 136–156). Adelaide, Australia: NCVER.

Davies, A., Fidler, D. and Gorbis, M. 2011. *Future work skills 2020*. Palo Alto, CA: University of Phoenix Research Institute.

Dawe, S. 2002. *Focussing on generic skills in training packages*. Adelaide, Australia: The National Council for Vocational Education Research.

Del Castillo, X.Á. 2000. *He trained in mechanical engineering, a vision of future* (Doctoral dissertation, Polytechnic University of Catalonia, UPC, Barcelona).

Department of Education and Skills. 2005. *About the department*. Retrieved from: http:www.dfes.gov.uk/aboutus/index/shtml [Accessed 16 March 2006].

DuBrin, A.J. 2007. *Leadership: Research findings, practice, and skills* (5th ed.). Boston, MA: Houghton Mifflin Company.

Durrani, N. and Tariq, V.N. 2012. The role of numeracy skills in graduate employability. *Education Training*, 54(5), pp. 419–434. http://dx.doi.org/10.1108/00400911211244704.

Eurydice. 2002. *Key competencies: A developing concept in general compulsory education*. Brussels, EU: Eurydice.

Evers, F.T., Rush, J.C. and Berdrow, I. 1998. *The bases of competence: Skills for lifelong learning and employability*. San Francisco, CA: Jossey-Bass Publishers.

Farooqui, R.U. and Ahmed, S.M. 2009. Key skills for graduating construction management students: A comparative study of industry and academic perspectives. *Proceedings of the Construction Research Congress 2009: Building a Sustainable Future*, American Society of Civil Engineers (ASCE), Seattle, WA, p. 1439.

Finch, D.J., Hamilton, L.K., Baldwin, R. and Zehner, M. 2013. An exploratory study of factors affecting undergraduate employability. *Education Training*, 55(7), pp. 681–704. http://dx.doi.org/10.1108/ET-07-2012-0077.

Fritz, S., Brown, F.W., Lunde, J.P. and Banset, E.A. 2005. *Interpersonal skills for leadership* (2nd ed.). Upper Saddle River, NJ: Pearson Education, Inc.

Furnell, J. and Scott, G. 2015. Are we all on the same page? Teacher, graduate and student perceptions of the importance of skills thought to enhance employability. *Journal of Learning Development in Higher Education*, 667(8), pp. 1–10.

Gardner, C.T., Milne, M.J., Stringer, C.P. and Whiting, R.H. 2005. Oral and written communication apprehension in accounting students: Curriculum impacts and impacts on academic performance. *Accounting Education*, 14(3), pp. 313–336.

Gbomita, V. 1997. The adoption of microcomputers for instruction: Implications for emerging instructional media implementation. *British Journal of Educational Technology*, 28(2), pp. 87–101.

Ghazali, G. and Bennett, D. 2017. Employability for music graduates: Malaysian educational reform and the focus on generic skills. *International Journal of Music Education*. doi: 10.1177/0255761416689844.

Gibbons-Wood, D. and Lange, T. 2000. Developing core skills: Lessons from Germany and Sweden. *Education and Training*, 42(1), pp. 24–32.

Graham, R., Crawley, E. and Mendelsohn, B.R. 2009. Engineering leadership education: A snapshot review of international good practice. *White paper sponsored by the Bernard Gordon-MIT Engineering Leadership Program*, Cambridge, MA, p. 41. Retrieved from: http://web.mit.edu/gordonelp/elewhitepaper.pdf [Accessed 30 November 2011].

Griffin, M. and Annulis, H. 2013. Employability skills in practice: The case of manufacturing education in Mississippi. *International Journal of Training and Development*, 17(3), pp. 221–232.

Ho, H.F. 2015. Matching university graduates' competences with employers' needs in Taiwan. *International Education Studies*, 8(4), pp. 122–133.

Honey, P. 2001. An identikit picture of a lifelong learner. *Training Journal*, p. 7.

Husain, M.Y., Mokhtar, S.B., Ahmad, A.A. and Mustapha, R. 2010. Importance of employability skills from employers' perspective. *Procedia: Social and Behavioral Sciences*, 7, pp. 430–438.

Jackson, D. 2013. Student perceptions of the importance of employability skill provision in business undergraduate programs. *Journal of Education for Business*, 88(5), pp. 271–279.

Jackson, D. and Chapman, E. 2012. Non-technical competencies in undergraduate business degree programs: Australian and UK perspectives. *Studies in Higher Education*, 37(5), pp. 541–567. doi: 10.1080/03075079.2010.527935.

Johari, M.H., Zaini, R.M., Zaharim, A., Basri, H. and Zaidi Omar, M. 2011. Perception and expectation toward engineering graduates by employers: A UKM study case. *Proceedings of the 3rd International Congress on Engineering Education (ICEED)*, Kuala Lumpur, Malaysia, pp. 203–207.

Jung, I. and Rha, I. 2001. A virtual university trial project: Its impact on higher education in South Korea. *Innovations in Education and Training International*, 38(1), pp. 31–41.

Keating, J., Medrich, E., Volkoff, V. and Perry, J. 2002. *Comparative study of vocational education and training systems: National vocational education and training systems across three regions under pressure of change.* Review of Research. Leabrook, South Australia: National Centre for Vocational Education Research.

Kelly, A. 2001. The evolution of key skills: Towards a Tawney paradigm. *Journal of Vocational Education and Training*, 53(1), pp. 21–35.

Kilgour, M. and Koslow, S. 2009. Why and how do creative thinking techniques work? Trading off originality and appropriateness to make more creative advertising. *Journal of the Academy of Marketing Science*, 37(3), pp. 298–309.

Lievens, F. and Sackett, P.R. 2012. The validity of interpersonal skills assessment via situational judgment tests for predicting academic success and job performance. *Journal of Applied Psychology*, 97(2), p. 460.

Lim, T., Fadzil, M., Latif, L.A., Goolamally, N.T. and Mansor, N. 2011. Producing graduates who meet employer expectations: Open and distance learning is a viable option. *Proceedings of International Conference on Languages, Literature and Linguistics (ICLLL)*, Kuala Lumpur, Malaysia.

Lim, Y.M., Lee, T.H., Yap, C.S. and Ling, C.C. 2016. Employability skills, personal qualities, and early employment problems of entry-level auditors: Perspectives from employers, lecturers, auditors, and students. *Journal of Education for Business*, 91(4), pp. 1–8.

Love, P., Haynes, N., Sohal, A., Chan, A. and Tam, C. 2002. *Key construction management skills for future success.* Monash University. Faculty of Business and Economics, Working Paper 49/02.

Mansour, B.E. and Dean, J.C. 2016. Employability skills as perceived by employers and university faculty in the fields of human resource development (HRD) for entry level graduate jobs. *Journal of Human Resource and Sustainability Studies*, 4(1), pp. 39–49.

Mat, N.H.N. and Zabidi, Z.N. 2010. Professionalism in practices: A preliminary study on Malaysian public universities. *International Journal of Business and Management*, 5(8), p. 138.

McLaughlin, M. 1992. *Employability skills profile: What are employers looking for?* Report 81–92-E. Ottawa, ON: Conference Board of Canada.

Messum, D., Wilkes, L. and Jackson, D. 2011. Employability skills: Essential requirements in health manager vacancy advertisements. *Asia Pacific Journal of Health Management*, 6(2), pp. 22–28.

Messum, D., Wilkes, L. and Jackson, D. 2015. What employability skills are required of new health managers? *Asia Pacific Journal of Health Management*, 10(1), pp. 28–35.

Messum, D., Wilkes, L.M., Jackson, D. and Peters, K. 2016b. Employability skills in health services management: Perceptions of recent graduates. *Asia Pacific Journal of Health Management*, 11(1), pp. 25–32.

Messum, D., Wilkes, L., Peters, K. and Jackson, D. 2016a. Content analysis of vacancy advertisements for employability skills: Challenges and opportunities for informing curriculum development. *Journal of Teaching and Learning for Graduate Employability*, 6(1), pp. 72–86.

Mihail, D.M. 2006. Internships at Greek universities: An exploratory study. *Journal of Workplace Learning*, 18(1), pp. 28–41. doi: 10.1108/13665620610641292.

Mirza, F.M., Jaffri, A.A. and Hashmi, M.S. 2014. *An assessment of industrial employment skill gaps among university graduates in the Gujrat-Sialkot-Gujranwala industrial cluster*, Pakistan.

PSSP working papers 17. Pakistan: International Food Policy Research Institute (IFPRI).

Misko, J. 2006. *Vocational education and training in Australia, the United Kingdom and Germany: Informing policy and practice in Australia's training system.* Adelaide, Australia: National Centre for Vocational Education Research.

Mitchell, G.W., Skinner, L.B. and White, B.J. 2010. Essential soft skills for success in the twenty-first century workforce as perceived by business educators. *The Journal of Research in Business Education*, 52(1), pp. 43–53.

Munap, R., Badrillah, M.I.M. and Mokhtar, A.R.M. 2015. Graduates' employability skills: Hard and soft skills towards employee productivity from the perspective of Malaysian employers. *Proceedings of the International Symposium on Research of Arts, Design and Humanities (ISRADH 2014)*, Springer, Singapore, pp. 151–158.

National Council for Vocational Education Research (NCVER). 2003. *Defining generic skills: At a glance.* Adelaide, South Australia: NCVER.

Ngowi, A.B., Pienaar, E., Talukhaba, A. and Mbachu, J. 2005. The globalisation of the construction industry: A review. *Building and Environment*, 40(1), pp. 135–141.

Nickson, D., Warhurst, C., Commander, J., Hurrell, S.A. and Cullen, A.M. 2012. Soft skills and employability: Evidence from UK retail. *Economic and Industrial Democracy*, 33(1), pp. 65–84.

Nowiński, W. and Haddoud, M.Y. 2019. The role of inspiring role models in enhancing entrepreneurial intention. *Journal of Business Research*, 96, pp. 183–193.

O'Leary, S. 2017. Graduates' experiences of, and attitudes towards, the inclusion of employability-related support in undergraduate degree programmes: Trends and variations by subject discipline and gender. *Journal of Education and Work*, 30(1), pp. 84–105.

Omar, N.H., Abdul Manaf, A., Helma Mohd, R., Che Kassim, A. and Abd Aziz, K. 2012. Graduates' employability skills based on current job demand through electronic advertisement. *Asian Social Science*, 8(9), pp. 103–110.

Overtoom, C. 2000. *Employability skills: An update.* Center on Education and Training for Employment, ERIC Digest No. 220, Report. Retrieved from: www.cete.org/acve/docgen. asp?tbl=digests&ID=105.

Packer, A.C. and Brainard, S. 2003. *Implementing SCANS. Highlight zone: Research@ Work.* Columbus, OH: National Dissemination Center for Career and Technical Education of The Ohio State University.

Pokras, S. 1995. *Team problem solving.* Menlo Park, CA: Crisp Publications, Inc.

Poon, J. 2012. Real estate graduates' employability skills: The perspective of human resource managers of surveying firms. *Property Management*, 30(5), pp. 416–434.

Pradhan, S. 2015. Study of employability and needed skills for LIS graduates. *DESIDOC Journal of Library and Information Technology*, 35(2), pp. 106–112.

Qomariyah, N., Savitri, T., Hadianto, T. and Claramita, M. 2016. Formulating employability skills for graduates of public health study program. *International Journal of Evaluation and Research in Education*, 5(1), pp. 22–31.

Rahmat, N., Ayub, A.R. and Buntat, Y. 2016. Employability skills constructs as job performance predictors for Malaysian polytechnic graduates: A qualitative study. *Malaysian Journal of Society and Space*, 12(3), pp. 154–167.

Rajapakse, R.P.C.R. 2017. Importance of soft skills on employability of finance graduates'. *Asian Journal of Multidisciplinary Studies*, 5(1), pp. 136–141.

Ramadi, E., Ramadi, S. and Nasr, K. 2015. Engineering graduates' skill sets in the MENA region: A gap analysis of industry expectations and satisfaction. *European Journal of Engineering Education*, 41(1), pp. 34–52.

Rawlins, J. and Marasini, R. 2011. Are the construction graduates on CIOB accredited degree courses meeting the skills required by the industry? *Management, 167*, p. 174.

Reid, J.R. and Anderson, P.R. 2012. Critical thinking in the business classroom. *Journal of Education for Business, 87*(1), pp. 52–59.

Richens, G. 1999. *Perceptions of southern Nevada employers regarding the importance of SCANS workplace basic skills* (Ph.D. dissertation, the University of Nevada, Las Vegas).

Robles, M.M. 2012. Executive perceptions of the top 10 soft skills needed in today's workplace. *Business Communication Quarterly, 75*(4), pp. 453–465.

Russell, J.S., Hanna, A., Bank, L.C. and Shapira, A. 2007. Education in construction engineering and management built on tradition: Blueprint for tomorrow. *Journal of Construction Engineering and Management, 133*(9), pp. 661–668. doi: 10.1061/ (ASCE)0733-9364(2007)133:9(661).

Rutherford, J. 2005. Key competencies in the New Zealand curriculum: Development through consultation. *Curriculum Matters, 1*(1), pp. 210–227.

Saad, M.S.M. and Majid, I.A. 2014. Employers' perceptions of important employability skills required from Malaysian engineering and information and communication (ICT) graduates. *Global Journal of Engineering Education, 16*(3), pp. 110–115.

Salleh, R., Md Yusof, M.A. and Memon, M.A. 2016. Attributes of graduate architects: An industry perspective. *The Social Sciences (Pakistan), 11*(5), pp. 551–556.

Samavedham, L. and Ragupathi, K. 2012. Facilitating 21st century skills in engineering students. *Journal of Engineering Education Transformations, 25*(4–1), pp. 37–49.

Scholte, J.A. 2008. Defining globalisation. *World Economy, 31*(11), pp. 1471–1502.

Schomburg, H. and Teichler, U. 2007. *Higher education and graduate employment in Europe: Results from graduates surveys from twelve countries* (Vol. 15). Dordrecht, Netherlands: Springer Science & Business Media.

Selvadurai, S., Choy, E.A. and Maros, M. 2012. Generic skills of prospective graduates from the employers' perspectives. *Asian Social Science, 8*(12), pp. 295–303.

Seth, D. and Seth, M. 2013. Do soft skills matter?: Implications for educators based on recruiters' perspective. *IUP Journal of Soft Skills, 7*(1), pp. 7–20.

Shiarella, A.H., McCarthy, A.M. and Tucker, M.L. 2000. Development and construct validity of scores on the community service attitudes scale. *Educational and Psychological Measurement, 60*(2), pp. 286–300.

Silverberg, M., Warner, E., Fong, M. and Goodwin, D. 2004. *National assessment of vocational education: Final report to Congress.* Washington, DC: US Department of Education, Office of the Under Secretary Policy and Program Studies Service.

Singh, R., Chawla, G., Agarwal, S. and Desai, A. 2017. Employability and innovation: Development of a scale. *International Journal of Innovation Science, 9*(1), pp. 20–37.

Srivastava, A. and Khare, M. 2012. *Skills for employability: South Asia.* National University of Educational Planning and Administration (NUEPA), New Delhi, India.

Steyn, Z. and Van Staden, L.J. 2018. Investigating selected self-management competencies of managers. *Acta Commercii, 18*(1), pp. 1–10.

Støren, L.A. and Aamodt, P.O. 2010. The quality of higher education and employability of graduates. *Quality in Higher Education, 16*(3), pp. 297–313.

Tatum, C. 2010. Construction engineering education: Need, content, learning approaches. In *Proceedings of the Construction Research Congress 2010: Innovation for Reshaping Construction Practice*, Construction Research Congress 2010, May 8–10, 2010, American Society of Civil Engineers (ASCE), Banff, AB, p. 183.

Turner, D. 2001. *Employability skills development in the United Kingdom.* Adelaide, South Australia: The National Centre for Vocational Education Research.

Turner, D. 2002. *Employability skills development in the United Kingdom*. Appendix. Kensington Park: National Centre for Vocational Education Research.

Valentine, S., Hollingworth, D. and Eidsness, B. 2014. Ethics-related selection and reduced ethical conflict as drivers of positive work attitudes: Delivering on employees' expectations for an ethical workplace. *Personnel Review*, 43(5), pp. 692–716.

Vashisht, A., Pandey, D.K. and Pathak, R. 2016. Education vs. employability: The need to bridge the skills gap among the engineering and management graduates in Madhya Pradesh (India). *Global Journal of Multidisciplinary Studies*, 5(5), pp. 72–76.

Warraich, N.F. and Ameen, K. 2011. Employability skills of LIS graduates in Pakistan: Needs and expectations. *Library Management*, 32(3), pp. 209–224.

Washer, P. 2007. Revisiting key skills: A practical framework for higher education. *Quality in Higher Education*, 13(1), pp. 57–67. doi: 10.1080/13538320701272755.

Wickramasinghe, V. and Perera, L. 2010. Graduates', university lecturers' and employers' perceptions towards employability skills. *Education+ Training*, 52(3), pp. 226–244.

Yang, H., Cheung, C. and Fang, C.C. 2015. An empirical study of hospitality employability skills: Perceptions of entry-level hotel staff in China. *Journal of Hospitality & Tourism Education*, 27(4), pp. 161–170.

Young, J. and Chapman, E. 2010. Generic competency frameworks: A brief historical overview. *Education Research and Perspectives*, 37(1), p. 1.

Yusuf, S., Saint, W. and Nabeshima, K. 2009. *Accelerating catch-up: Tertiary education for growth in Sub-Saharan Africa*. Washington, DC: World Bank.

Yuzainee, M.Y., Zaharim, A. and Omar, M.Z. 2011. Employability skills for an entry-level engineer as seen by Malaysian employers. *Proceedings of the IEEE 2011: Global Engineering Education Conference (EDUCON) 2011*, Amman, Jordan, pp. 80–85.

6 Overview of South African education

Overview of South African education

Nearly two decades after the end of the apartheid era, the labour market in South Africa continues to be plagued by persistent levels of unskilled labour, unemployment, substantial inequality in earnings, and skill shortages in relevant areas (Scott *et al.*, 2007). According to Branson and Zuze (2012), 'the country's adverse labour market outcomes are rooted in an ailing schooling system where access to quality education is still inequitably distributed along the lines of race and socio-economic background'. The inability of certain groups, particularly those historically disadvantaged, to access quality education means that the opportunity to participate in emerging economic opportunities is effectively denied to a large section of the citizenry. This implies that inequalities in the education system predictably generate labour market inequalities, a situation that hinders social transformation and thwarts long-term economic development goals of the nation.

Considering the various socio-economic challenges facing South Africa, the role of universities in delivering quality and responsive graduates to enhance the nation's economy cannot be overemphasized. Therefore, it is germane to comprehend higher educational outcomes, what they have provided, and how the system intends to provide resources to meet the growing labour market needs (Fisher and Scott, 2011; Filmer, 2012). Today, the development of students is dependent on the educational processes they undergo during their university education. Graduates of today are regarded as the future of tomorrow and custodians of the construction industry as they are poised to take up positions in design and supervisory roles. Hence, the principal aim of universities is to provide students with a holistic blend of education that will equip them with the required skills to thrive after graduation. Several evolutions and transformations have occurred in every aspect of the South African society as the quest for a true democracy in the post-apartheid era continues. One important sector where these evolutions and transformations have occurred and are still occurring is in the higher education sector (Soni, 1998). The sector is viewed as one of the most significant in creating an intellectual workforce (citizenry) as the nation strives for a knowledge-based system. These views are comprehensively stated in the various policy documents relating to education across the nation.

The higher education sector is also confronted by several challenges as reported during a seminar in July 2002 hosted by the Centre for Higher Education Transformation (CHET), the Committee of Technikon Principals (CTP), the South African Universities Vice Chancellors Association (SAUVCA), and UNITECH (Universities and technikons communication practitioners). The seminar on Entrepreneurial Higher Education Institutions noted that universities are no longer the sole creators and disseminators of knowledge owing to the advent of other knowledge generators, including business firms, independent think tanks, and knowledge laboratories. Findings from the seminar also revealed that institutions face challenges from students and employers as well as other education providers. Through the introduction of education providers with strong commercial and vocational inclinations, a variety of higher education institutions exist, thereby increasing the options for students. Moreover, the influence of information and communication technology (ICT) may have a dramatic bearing on the functioning of higher education systems.

Several key legislations and policies governing education in South Africa are discussed next.

Legislation and policies developments post-1994

Academic programmes for higher education are designed to deliberately achieve learning outcomes for the benefit of its citizenry. These programmes are instrumental in promoting lifelong learning, providing additional education and training, and offering learning outcomes (specific and critical cross-field). Hence, it is germane to study the various legislation and bodies that have some level of influence in the education and training process.

In 1996, The National Education Policy Act (NEPA) was constituted and it established the policies and monitoring duties of the Minister of Education. It spearheaded the collaboration of intergovernmental forums such as the Council of Education Ministers (CEM) and the Heads of Education Departments Committee (HEDCOM) in ensuring a more developed education system. The NEPA also led to the advent of national policies, including Technical and Vocational Education and Training (TVET), to enhance curriculum and quality assurance. In the same year, the South African Schools Act (SASA) was formed. The Act aimed at ensuring the accessibility of quality education to all individuals (compulsory for learners 7–15 years of age) without any form of discrimination. The Education Laws Amendment Act amended this Act in 2005. The new Act states that schools in poverty-ridden areas are to be declared 'no-fee schools'. The Education Laws Amendment Act 2007 further modified it to address matters relating to school principals.

In 1998, the Employment of Educators Act was formed. This Act focused on ensuring the competence of educators or teachers with emphasis on their moral, ethical, and professional responsibilities. Concurrently, this Act works in tandem with the South African Council for Educators (SACE) in the regulation of teaching standards and qualifications.

In 2000, the Adult Basic Education and Training (ABET) was formed. ABET fosters the support, registration, and development of both private and public adult learning centres. In 2001, the Education White Paper 6 on Inclusive Education states the intention of the Department of Basic Education (DBE) to enforce inclusive education at all levels by 2020. The system caters for vulnerable learners and reduces learning barriers by developing structures and mechanisms to provide effective learning, particularly for those who are susceptible to dropping out owing to several reasons. After assuming power in the early days of the post-apartheid era, the African National Congress (ANC) implemented its mandate to establish an education and training system by appointing several commissions to deal with various aspects of the education system.

White Paper on Education and Training

Established in 1995, the White Paper on Education and Training was the first policy document to announce the new age of education and training in South Africa. The then Minister of Education, Prof. Emmanuel Bengu, echoed in a document the major challenges facing education and training: 'South Africa has never had a truly national system of education and training' (Van Wyk and Mothata, 1998). This policy document officially proposed and implemented an integrated education and training system while proposing several key values and principles as described in the Bill of Rights of the New Constitution.

An integrated approach to education and training

This document identifies education and training as a joint unit rather than an unorthodox division in the past among the academic space. Both of them are spheres where skills, knowledge, and competencies (both academic and non-academic) are acquired. Hence, both education and training are key elements of human capital development and should be treated as symbiotic activities. Given the fact, the White Paper on Education and Training supports an integrated approach to education and training, a trend that is practised globally in curriculum development and qualification reforms. Consequently, the Ministers of Education and Labour proposed an Inter-Ministerial Working Group to foster the integrated approach in education and training. This group argued that both education and training should work together and be seen as a lifelong process (Department of Education, 1997a). This is further clarified when discussing the National Qualification Framework (NQF) later on.

Another critical element of the integrated approach to education and training is the value placed on the nature of learning that individuals have already acquired. The acknowledgement of what individuals already know is known as the Recognition of Prior Learning (RPL), one of the foundations of the NQF. Therefore, the integrated approach infers that individuals can access the education and training system at a certain level contingent on their previous knowledge or learning. For example, a man who dropped out of school aged 14 with a Grade

8 and then hopes to resume school aged 40 may not necessarily commence at Grade 9. He will be assessed in terms of skill base and current knowledge and then reintroduced into the system at the level that suits him best (Human Science Research Council [HSRC], 1995). Similarly, the integrated approach allows individuals or learners to continue level progressions by accruing credit units for stages they have successfully completed. For example, a 40-year-old learner can use his credit units to take up positions in the world of work. In this case, the credit units are transferable or convertible across different schemes or programmes, education providers, and industries. This is achievable through the NQF that is responsible for unifying qualifications in education and training that are contingent according to the standards and assessment procedures (HSRC, 1995).

Lifelong learning

The White Paper on Education and Training also provides for lifelong learning that involves individuals seeking continuous learning and striving for adaptability to improve themselves and knowledge development goals (Van Wyk and Mothata, 1998). Apart from the provision of educational prospects, lifelong learning also involves creating learning opportunities and providing individuals with the choice to learn when, how, and where they want it to happen.

Outcome-based learning approach

The integrated approach to education and training and the lifelong learning concept are rooted in the outcome-based education (OBE) and the NQF. This learning system is possible for people whose career paths or academic progress have been hampered by failing to undergo the necessary evaluation or certification of their previous knowledge or qualifications (Van Wyk and Mothata, 1998).

Independent and critical thought

The White Paper on Education and Training also proposed that the teaching methods and curricula should encourage independent and critical thinking abilities. This involves the ability and capacities to communicate effectively, carry out reasonings, make inquiries, and form judgements. These outlined capacities are rooted in the outcome-based approach.

Mathematics, science, and technology initiative

The White Paper on Education and Training places premium emphasis on mathematics, science, and technology because they are vital for economic progress and societal stability. Comparatively speaking, Van Wyk and Mothata (1998) suggest that South African learners struggle to excel in mathematics and science compared to learners from more advanced nations of the world. This stemmed from the apartheid education that prevented learners from excelling in these disciplines. Hence, focusing on the enhancement of learners in the areas of mathematics and science is considered essential to socio-economic success.

Changing legacies of the past

The White Paper on Education and Training advocates for the provision of excellent learning opportunities and educational prospects for all learners, irrespective of race, gender, or orientation (Van Wyk and Mothata, 1998). This was done because of the previous underdevelopment and inequity that plagued the educational sector in the past.

Seamless accessibility to education and training

The NQF's roles as discussed later suggest that learners have the opportunity to learn across various learning contexts, a prospect that enhances the lifelong learning process (Van Wyk and Mothata, 1998; Ngcongo and Chetty, 2000). The White Paper on Education and Training also advocates equity and implies that the citizenry experiences the same quality of learning opportunities. The policy advocates the adoption of several approaches in ensuring that both learners and teachers are offered quality treatment in the educational sphere and beyond. Consequently, the Employment Equity Bill was proposed to eradicate unfair discrimination and imbalances in the employment sector (Ngcongo and Chetty, 2000).

Parents' rights

The White Paper on Education and Training also advocates for parents and guardians to take major responsibility for the educational development of their children and can provide advice to state authorities on how their education can be improved (Ngcongo and Chetty, 2000).

Rehabilitation of schools

The White Paper on Education and Training addresses the rehabilitation of educational centres, including colleges and schools. It implies that rehabilitation must work in tandem with restoring the ownership of institutions to their communities via the formation of legitimate governing bodies.

Accountability

The White Paper on Education and Training also calls for the creation of an accountability culture in the areas of teaching and learning. This calls for various stakeholders, including learners, teachers, and even the government, to establish a common purpose or mission in executing their duties effectively. This involves a consensual understanding of the various responsibilities required to thrive.

Cultural, language, and religious traditions

Also addressed in the White Paper on Education and Training is the aspect of people's cultural differences, language, and religions. This is because learning correlates with the outside world of the individual, which is one of the cornerstones of outcomes-based education.

Based on the key values and principles already discussed, the White Paper on Education brought about change by establishing a set of proposals to reform the connection that exists between education and training.

South African Qualifications Authority

Another key policy that was formed was the South African Qualification Framework (SAQA). To understand the roles of SAQA, its design structure, general operations, and functionalities are discussed next.

Structure of South African Qualifications Authority

Regarded as the qualification accrediting body in South Africa, SAQA was established in 1995 and involves up to 28 members. This includes a chair and up to 6 other members appointed by the Minister of Education and 21 further members elected for appointment by different stakeholders (Vakalisa, 2000). The stakeholders who are eligible for electing SAQA representatives include councils of the university, directors-general of education and labour, the National Training Board (NTB), nationwide groups representing organized labour and commerce, and provincial education departmental heads. Therefore, SAQA is a large body that comprises a wide range of both providers and consumers of the nation's education and training resources (Vakalisa, 2000). This wide range of stakeholders ensures that the body is continually equipped with the right expertise needed for the assessment of qualification standards, a role SAQA is primarily designed to execute.

Functions of SAQA

As described earlier, SAQA's principal role is to ensure and supervise the implementation of the NQF, which includes 'establishing education and training standards or qualifications and accrediting bodies responsible for monitoring and auditing achievements in terms of such standards and qualifications'. SAQA is also responsible for 'registration of accreditation of bodies responsible for establishing education and training standards of qualifications and assigning function to those bodies, registration of national standards; ensuring compliance with provisions for accreditation and ensuring that standards and registered qualifications are internationally comparable'. SAQA also works in tandem with the Minister of Education on issues regarding the registration of qualifications and standards.

SAQA and subsidiary bodies

The mission of SAQA is to 'to develop and sustain policies, procedures, and infrastructure for the National Qualifications Framework'. As a large body, SAQA fulfils this mission through four structures. These are briefly described.

National standards bodies

The authority of South African Qualifications Authority includes determining the NSBs' size whose functions include 'advising SAQA on the sub-fields which constitute their field of learning and setting up the Standards Generating Bodies (SGBs) for each sub-field'. Other functions include

> examining qualification proposals made by education providers and recommending the registration of unit standards of programmes that meet their approval on the NQF and monitoring the provisioning of educational programmes within their field of expertise through the well-defined functions of the Education and Training Quality Assurers (ETQAs).

Succinctly, the NSBs work in tandem with SAQA by guiding the body on the acceptable standards of qualifications both nationally and globally (Vakalisa, 2000).

Standards generating bodies

As stated earlier, one of the NSB's roles is the formation of SGBs in sub-fields. Moreover, individuals who are appointed to function in SGBs are specialists in their respective fields of influence and control. The SGBs are represented by a wide spectrum of stakeholders, including education and training providers, community groups, state departments, and organized labour (Vakalisa, 2000). The functions of the SGBs include 'generating unit standards and qualifications in accordance with SAQA requirements in identified sub-field and levels and updating and reviewing standards, recommending unit standards and qualifications in NSBs' (Vakalisa, 2000).

Education and training quality assurers

The Education and Training Quality Assurers (ETQAs) ensure compliance based on the standards set by SAQA regarding the quality of education and training. A key function of the SGB's is the

> accreditation of constituent providers requiring the evaluation of the quality management system of a provider to see whether it can demonstrate the ability to provide learning programmes and manage the assessment of those qualifications and, or unit standards for which it wishes to be accredited.

Other functions of the ETQAs include the 'promotion of quality among constituent providers, monitoring the manner in which the programmes are presented by constituent providers, registration of constituent assessors for specified NQF standards and or qualifications in terms of the criteria established for that particular purpose'. The ETQAs are also 'responsible for the certification of learners, recommending new standards and qualifications to NSBs for consideration or the recommendation of modifications to existing NQF standards and qualifications to NSBs for consideration and any other function assigned to them by SAQA'.

The qualifications councils

Qualification councils combine qualifications in different learning programmes to determine individuals' eligibility to receive certificates upon completion of the three education and training stages as specified in the NQF. These three education and training bands are the General, Further, and Higher Education. The qualification councils also work in tandem with the relevant stakeholders, including academic and professional bodies, and are responsible for the NQF qualification registration (HSRC, 1995; Vakalisa, 2000).

National qualifications framework

The NQF was established to improve the quality of education in South Africa. In accordance with the nationwide qualification levels, the NQF provides lifelong learning opportunities for individuals irrespective of age, gender, and education level. It consists of 'standards and qualifications that can be formally or informally obtained across the nation' (Vakalisa, 2000).

Principles supporting the NQF

Several principles underpin the NQF which were established in the National Training Strategy Initiative process (Mothata, 1998). They include integration that 'forms part of human resources development and provides for the establishment of a unifying approach to education and training' and recognition of prior learning which 'gives credit to learning which has already been acquired in different ways, e.g. through life experience'. Other principles include access that provides 'easy access to appropriate levels of education and training to all learners in a manner that facilitates progression'; coherence that deals with 'working within a consistent framework of principles and certification', and legitimacy that 'provides for the participation of all national stakeholders in the planning and co-ordination of standards and qualifications'. In addition, portability 'enable[s] learners to transfer credits and qualifications from one learning situation or employer to another'; credibility 'deals with national and international acceptance (in terms of qualifications)'; standards 'deal with a nationally agreed framework and internationally acceptable outcomes', and flexibility 'allow[s] for multiple pathways to the same learning ends'.

There is also progression that 'ensures that the framework of qualifications permits individuals to move through the levels of national qualifications via different appropriate combinations of the components of the delivery system' and guidance of learners that 'provide[s] for the counselling of learners by specially trained individuals who meet nationally recognised standards for educators and trainers'. Based on these principles, the NQF is seen as an avenue for education and training system restructuring. However, this framework can become more effective by adopting an outcomes-based approach instead of the conventionally used content-based approach. Furthermore, the White Paper on Education and Training asserts that the NQF 'encourages the creation of new and flexible curricular, promote the upgrading of learning standards, monitor and regulate the quality of qualifications and permit a high level of articulation between qualifications based on the recognition and accumulation of credits'.

Objectives of the NQF

According to SAQA, there are several objectives of the NQF. They include the

creation of an integrated national framework for learning achievements, facilitating access to, and mobility and progression within the education, training and career paths, enhancing the quality of education and training and accelerating the redress of past unfair discrimination in education, training and employment opportunities.

The NQF also 'contribute[s], thereby, to the full personal development of each learner, and the social and economic development of the nation at large' (Mothata, 1998).

Structure of the NQF

In meeting the needs of the already mentioned objectives, the NQF summarizes the various national qualifications and standards 'that guide accreditation of learning qualifications that are obtained through the participation in any learning programme' (Vakalisa, 2000). The NQF shows the different paths of training and formal education that learners take in acquiring qualifications from the lowest at level 1 to the highest at level 8.

The introduction of Curriculum 2005

In 1997, Prof. Bengu, the then Minister of Education, announced the Curriculum 2005. This new curriculum supports the idea of an integrated system, underscores lifelong learning, and backs the NQF that comprises three levels of education and training (General, Further, and Higher education) (Kraak and Hall, 1999). Curriculum 2005 was identified as a pivotal process in transforming the South African society by fostering the integration of education and training through the NQF. One of its key mandates was to achieve a thriving, democratic, united and internationally competitive nation with a knowledgeable, creative, and productive citizenry, living in a country free from prejudice, unfairness, violence, and discrimination (Department of Education, 1997b).

Structure and design of Curriculum 2005

The following sections describe the various structures and design features of the Curriculum 2005.

Critical cross-field education and training outcomes

Under the supervision of SAQA, the various stakeholders in the education and training sectors developed the critical cross-field education and training outcomes. Directly and indirectly, it is the responsibility of the critical cross-field outcomes to maintain the NQF. These critical outcomes are the required results that both educators and learners involved in teaching and learning should possess at every developmental level. For learners, it refers to the various skills, knowledge, values, and abilities needed to be successful across various contexts and to affect their communities and immediate society positively (Rensburg, 1998). According to SAQA (1997), critical outcomes direct the teaching process to help learners by 'identifying and solving problems in which responses show that responsible decisions, using critical and creative thinking, have been made, working effectively with others as a member of a team, group, organisation, community, and collecting, analysing, organising and critically evaluating information'. The critical outcomes also help learners to

> communicate effectively using visual, mathematical and or language skills in the modes of oral and written presentation, using science and technology effectively and critically and demonstrating an understanding of the world as a set of related systems by recognizing that problem-solving contexts do not exist in isolation.
>
> (SAQA, 1997)

Specific outcomes

As previously discussed, specific outcomes refer to the various skills, knowledge, values, and abilities needed to be successful across various contexts. Rensburg (1998) suggests that these outcomes are the foundations upon which learners can function effectively in a given field at any given point in time. Moreover, these specific outcomes were produced for the eight learning areas identified on the NQF and indicate the outcome attained at the end of the learning process (Rensburg, 1998). The specific outcomes also clarify 'the importance of the learning area and why it should be included in the curriculum, the essential elements of the learning area, how the learning area will contribute to the achievement of the critical outcomes' (SAQA, 1997).

Learning areas

Learning areas are the various spheres through which learners peculiar to the General Education and Training (GET) benefit from a balanced curriculum. They also provide a basis for the development and implementation of learning programmes and curricula (Rensburg, 1998). There are eight different learning areas. These include 'Arts and Culture, Economic and Management Sciences, Human and Social Sciences, Language Literacy and Communication, Life Orientation, Mathematical literacy, and its Sciences, Natural Sciences and Technology' (Department of Education, 1997c).

Learning programmes

A clear difference exists between learning areas and learning programmes. According to Rensburg (1998), the eight different learning areas mentioned previously pave the way for the development of learning programmes that underpin learning. Learning programmes are considered as the vehicles used to implement curricula in achieving specific outcomes. Furthermore, learning programmes encapsulates the following design characteristics: assessment criteria, critical outcomes, performance indicators, range statements, and specific outcomes (Rensburg, 1998). From the aforementioned, learning programmes are seen as different sets of learning activities.

Implementation plan for Curriculum 2005

As early as 1995, the then Minister of Education, Prof. Bengu, pioneered the new curriculum and planned full implementation from Grades 1 through 12 by 2000. Nevertheless, the Council of Education Ministers integrated the OBE across both General and Further Education levels in 1997. This implementation strategy developed a paradigm that was to be launched in Grade 1 in 1998, leading to the advent of the Further Education and Training Certificate (FETC) and was awarded to the Grade 12 class of learners in 2005. This prompted the 2002 Curriculum Review Committee Report to coin the popular name 'C2005'. However, it is reported that the implementation process was hindered by several difficulties. The Curriculum Review Committee Report (2003) states that 'despite enormous political will and effort, social demands were seemingly not matched by financial, physical and human capacity in the system to implement the curriculum according to schedule'. The report also confirmed that the implementation of C2005 in the first grade of the senior phase; that is Grade 7, was postponed from 1999 to 2000. Pilots in Grades 3 and 7 began in 1999 and full implementation took place in the year 2000. In the year 2000, a pilot study for Grades 4 and 8 also took place, although all provinces did not participate in the study as noted by the Curriculum Review Committee Report (2003).

National Commission on Higher Education (NCHE)

Shortly after assuming power in 1994, the government appointed this commission to analyse and make recommendations on higher education matters. Dr. Jairam Reddy, who was a former Vice Chancellor of the University of Durban-Westville, chaired the commission and filed its report titled A Framework for Transformation in 1996. In several ways, the NCHE was instrumental in setting the tone for South African higher education. The commission stressed the need for transformation and revisited the principles underpinning a new higher education system. The NCHE had proposed several key recommendations, most of which were either implemented or are being implemented. Three envisaged pillars for a transformed higher education system underpin the successful implementation of the NCHE report. The first states that

> to satisfy the need of equity, redress, and development, a policy of increased participation was required. This should be achieved, in the NCHE's view, through a change from an elite higher education system to that of a mass higher education system (a process of massification).

Secondly, 'A policy of greater responsiveness was needed to ensure that higher education engaged with the challenges of its social context'. The third policy pillar suggests that 'increased cooperation and partnerships led to the recommendation of a model of 'cooperative governance', whose elements included the state in a supervisory role (as opposed to control or interference)'.

White Paper 3: a programme for higher education transformation

Following the 1996 NCHE report, the next year saw the release of the White Paper 3 – A Higher Education Transformation Program of the Department of Education. The White Paper emphasized the recommendations contained in the NCHE report as proposing them as a government policy. The White Paper deals with 'the functioning of the NQF with respect to higher education qualifications and the importance of quality control and the functioning of the HEQC'.

The Higher Education Act

After the White Paper 3, the Higher Education Act of 1997 was adopted. This Act initially provided the legal foundation for the adoption of policies developed and implemented by the NHCE. The Act contains several chapters that deal with certain aspects of the higher education setting. Some of these aspects include the level of autonomy to be enjoyed by institutions, the level of autonomy to be enjoyed by public higher education institutions, and the level of power to be wielded by the Minister of Education.

Structure of South African higher education sector

The South African higher education sector comprises the Minister of Education, who heads the Department of Education (DoE), and other bodies are as follows:

Council on Higher Education

The South African Council on Higher Education was formed in May 1998 in accordance with the Higher Education Act, No. 101 of 1997. As an independent statutory body, it works in tandem with the Minister of Education in addressing higher education matters such as policy issues, quality assurance, and training.

South African Universities Vice Chancellors Association

Established by Sections 6 and 7 of the Universities Act (1955), the Committee of University Principals (CUP) was initially a statutory body. After a series of restructuring and transformation, the committee is presently known as SAU-VCA. This association works in tandem with the Minister and Director-General of Education on general issues that are considered pertinent to universities. The body is also responsible for the appointment or nomination of qualified individuals to positions in statutory committees and councils that are of interest to universities. It is the duty of SAUVCA to formulate Joint Statutes and Joint Regulations that benefit several university policies and provisions which involve interrelations with other universities. With its role in regulating academic standards and engaging in policy matters, the functions of SAUVCA are hugely beneficial to the nation's public universities. SAUVCA is also responsible for the regulation of the Matriculation Board (MB) through which the body advises the Minister on matriculation issues that are basic requirements for enrolling for a university degree. SAUVCA fulfils its mandates by engaging in policy and legislation debates, participating and hosting seminars and symposiums, and paper presentations at conferences. Some specialist committees currently serve the SAUVCA to effectively carry out their duties. These are the Education Committee, Equity Committee, Executive Committee, Finance Committee, Intellectual, and the Legal Committee.

Centre for Higher Education Transformation

The Centre for Higher Education Transformation (CHET) is tasked with fostering transdisciplinary skills for project execution by scouting knowledgeable individuals from both the national and international higher education sectors. The CHET uniquely responds to the needs of higher education through a flexible system of operations that are manned by consultants and steering committees. The platform created by CHET facilitates the interaction between different stakeholders and higher education structures. This function has enabled CHET to collaborate with several local committees, including the Committee of College

Education Rectors South Africa, the Committee of Technikon Principals, the Committee of University Principals, the Ministry of Education, and the National Centre for Student Leadership. On the international front, CHET has been able to join forces with the American Council on Education, the Association for African Universities, the Centre for Higher Education Policy in the Netherlands, and the Commonwealth Higher Education Management Services.

Higher Education Quality Committee

Launched in 2001, the Higher Education Quality Committee (HEQC) is a permanent subcommittee of the Council on Higher Education that deals with quality assurance-related activities in the higher education sector. As stated in its mission and vision statements, the HEQC enhances the quality of both private and public institutions that would create an effective system for improved higher education and training among stakeholders involved. In accordance with the SAQA Act of 1995 (Act No. 58 of 1995) and the SAQA Regulations of 1998, the HEQC oversees the accreditation of the learning programmes of private and public institutions. Therefore, the main objective of the HEQC is to ensure that institutions deliver quality education and training, both efficiently and effectively, to develop skilled graduates who are responsive to the needs of society. Their duty also involves auditing the quality assurance compliance of higher education institutions and accrediting viable programmes of higher education. 'Quality' as defined by the HEQC guidelines is the fitness of the purpose of education providers regarding national goals and specified missions within a national framework. It is also viewed as the value for money in relation to the effectiveness and efficiency of providing quality education to benefit students as set out in the White Paper on Higher Education.

South African Qualifications Authority

The South African Qualifications Authority (SAQA) is a body comprising 29 members who are nominated by national stakeholders in the education and training sector and are appointed by the Ministers of Education and Labour. One of the key functions of SAQA is to oversee the development of the National Qualification Framework (NQF) by formulating and enforcing policies for the registration of bodies which are responsible for the establishment of education and training qualifications and standards. SAQA also oversees the enactment of the NQF regulations and principles, the registration of its national standards and qualifications, and the accreditation process of bodies tasked with the monitoring and auditing. Both nationally and globally, SAQA is recognized for its role in upholding the tenets of the NQF in terms of qualifications and part-qualifications. The Authority also conducts research and serves as the national source of recommending and evaluating domestic and foreign learning and qualifications.

National Qualifications Framework

The National Qualifications Framework (NQF) is a charter consisting of guidelines and principles that provide a vision and an organizational struc-ture to ensure the smooth running of a qualification system. Grounded on the principles of outcomes-based education, the NQF represents a national effort to ensure that education and training are blended into recognized qualifica-tions. The NQF takes into consideration the achievement of learners to high-light acquired knowledge and skills, thereby ensuring a system that fosters lifelong learning. According to the NQF, learning is arranged into 12 fields that are subdivided into several sub-fields. Subsequently, SAQA formed 12 National Standards Bodies (NSB) to address each of the organizing fields. The members who constitute the NSBs are acquired from the six elector-ates: community organizations, critical interest groups, education and train-ing providers, organized business, organized labour, and state departments. The NSBs are responsible for sanctioning qualifications and standards for the NQF to SAQA.

The new Higher Education Qualification Framework

The NQF's revised policy in the HEQF mould increases the NQF's efficiency to be more responsive to national needs. More specifically, the HEQF aims to support citizens who have historically been left out of national education and training systems so that they can realize their potential and skills. Therefore, this revised policy or framework was designed to tackle the challenges facing higher education in the 21st century. This policy also monitors higher education activities by supervising the development of programmes and qualifications that equip graduates with the necessary skills and competencies to foster economic transformation and social growth. This is because the role of higher education is critical in determining the socio-economic development of modern societies. Moreover, higher education institutions in South Africa, through the DoE, are accountable to the government and must register their qualifications through SAQA on the HEQF.

Historical antecedent of South African universities

Established in 1873, the first South African university was called the University of the Cape of Good Hope. The establishment of the maiden university resulted in two colleges being formed in 1829 and 1865, namely the South African College in Cape Town and the Victoria College in Stellenbosch, respectively. Rhodes University was next in line and was established in 1904 before both the South African College and Victoria College changed their names in 1918 to the Universities of Cape Town and Stellenbosch, respectively. Subsequently, the University of the Cape of Good Hope became the University of South Africa. The South African Native College was formed in 1916 by missionaries and later became the University of Fort Hare in 1951. In addition, the School of Mines was formed in 1895 and later became the University of the Witwatersrand in 1922.

Initially, the University of South Africa was a conglomerate of several university colleges before many of them became fully fledged universities. In 1959, the extension of the University Education Act that was designed to prohibit black students from attending historically white institutions led to the formation of racially segregated universities. These led to the advent of institutions such as the University of Durban-Westville, the University of the Western Cape, and the University of Zululand. Between the mid-1960s and mid-1980s, the Medical University of South Africa, the Rand Afrikaans University, the University of Port Elizabeth, and Vista University were established (Ferreira, 2003).

Furthermore, the South African educational system consists of three different components: primary or general education and training (GET), further education and training (FET), and higher education and training (HET).

Summary

This chapter provided an overview of South Africa's educational system and various legislations and policies that have some degree of influence in the education and training process. Also presented was a brief historical background of South African universities and the various components of the South African educational system. This chapter was important in understanding higher education outcomes, what they provided, and how the system intends to provide resources to meet the labour market's growing needs. The next chapter discusses the conceptual framework underpinning this study.

References

Branson, N. and Zuze, T.L. 2012. Education, the great equaliser: Improving access to quality education. *South African Child Gauge*, pp. 69–74.
Department of Education. 1997a. *Curriculum 2005. Life-long learning for the 21st century.* Pretoria: Department of Education.
Department of Education. 1997b. *Foundation policy phase (Grades 1–3).* Pretoria: Government Printers.
Department of Education. 1997c. *Intermediate phase policy document (Grades 4–6).* Pretoria: Government Printers.
Ferreira, M. 2003. *A framework for continuous improvement in the South African higher education sector* (Doctoral dissertation, University of Pretoria, Pretoria).
Filmer, D. 2012. Challenges and options for technical and post-basic education in South Africa. In *Policy Note, World Bank, Africa Region.* Washington, DC: Human Development Unit.
Fisher, G. and Scott, I. 2011. Background paper 3: The role of higher education in closing the skills gap in South Africa. *Closing the skills and technology gap in South Africa.*
Human Science Research Council (HSRC). 1995. *Ways of seeing the National Qualification Framework.* Pretoria: HSRC.
Kraak, A. and Hall, G. 1999. *Transforming further education and training in South Africa: A case study of technical colleges in KwaZulu-Natal. Volume one: Qualitative findings and analysis.*

Mothata, M. 1998. The National Qualifications Framework (NQF). In F. Pretorius (Ed.), *Outcomes-based education in South Africa*. Randburg, South Africa: Hodder and Stoughton Educational.

Ngcongo, R. and Chetty, K. 2000. Issues in school management and governance. In M. Thobeka and S. Mothata (Eds.), *Critical issues in South African education–after 1994*. Cape Town: Juta and Co.

Rensburg, J. 1998. Outcomes-based teaching and learning: Concepts and essentials. In F. Pretorius (Ed.), *Outcomes-based education in South Africa*. Randburg, South Africa: Hodder and Stoughton Educational.

Scott, I., Yeld, N. and Hendry, J. 2007. A case for improving teaching and learning in South African higher education. *Higher Education Monitor*, 6(2), pp. 1–8.

Soni, T.D. 1998. *Teaching strategies in higher education for the promotion of critical reflection* (Doctoral dissertation, University of Johannesburg, Johannesburg).

South African Qualifications Authority (SAQA). 1997. *The South African qualifications authority bulletin*. May/June (Vol. 1(i)). Pretoria: Office of the Executive Office SAQA.

Vakalisa, N. 2000. Lifelong learning. In M. Thobeka and S. Mothata (Eds.), *Critical issues in South African education–after 1994*. Cape Town: Juta and Co.

Van Wyk, N. and Mothata, S. 1998. Developments in South African education since 1998. In F. Pretorius (Ed.), *Outcomes-based education in South Africa*. Randburg, South Africa: Hodder and Stoughton Educational.

Section III

Conceptual framework underpinning employability

7   Conceptual framework underpinning employability

Selection of variables for improved employability

A conceptual framework is defined as a system of concepts, beliefs, expectations, assumptions, and theories that support one's research (Maxwell, 2005). It is also described as a written or visual product that expounds on the key factors, variables, or concepts in a narrative form or graphical representation and the supposed relationships among them (Miles and Huberman, 1994). They further suggest that one of the aims of utilizing a conceptual framework is to provide necessary direction on how the empirical study can be conducted or undertaken. Conceptual frameworks are constructed on the basis of the researcher's perception of the phenomenon. The adoption of existing models or theories helps to refine the researcher's goals, develop relevant research questions, choose suitable methods, and identify potential validity threats to the deductions of the study.

Nine frameworks (models) of employability were critically reviewed earlier in this book. However, Dacre Pool and Sewell's CareerEDGE model provided several constructs for employability which are adopted by this book. One of the strengths of the CareerEDGE model is that it improves on other existing models such as the USEM and DOTS model. This model was developed to practically explain the concept of employability, which has been described as indefinable, particularly to students and their parents (Dacre Pool and Sewell, 2007). The acronym 'Career-EDGE' was developed for easy recall: it means that, to be employable, all five elements (Career development learning, Experience – work and life, Degree subject knowledge, understanding, and skills, Generic skills, and Emotional intelligence) need to be developed. This will result in the development of self-confidence, self-efficacy, and self-esteem, which are key attributes for employability. This model is premised on the definition that 'employability is having the set of skills, knowledge, understanding and personal attributes that makes a person more likely to choose and secure occupations in which they can be satisfied and successful' (Dacre Pool and Sewell, 2007). Dacre Pool and Sewell contend that this model combines the earlier works by Hillage and Pollard (1998), Harvey (2002) and Yorke and Knight (2006) into a more logical model, which highlights the essential issues that support the concept of employability. The integration of several components of the USEM into the CareerEDGE model reflects the importance of the model.

This book adopts the critical components of the CareerEDGE model, namely Generic Skills (GS), Discipline-Specific Skills (DSS), Work-Integrated Learning (WIL), and Emotional Intelligence (EI). This book therefore considers Improved Graduate Employability (IGE) for built environment graduates as including Generic skills (GS) with 16 variables; Discipline-Specific Skills (DSS) with 12 variables; Work-Integrated Learning (WIL) with 21 variables; and Emotional Intelligence (EI) with 14 variables. However, the framework for this book includes University–Industry Collaboration (UIC) with 16 variables and Industry 4.0 (4IR) knowledge with 19 variables. Both additions are the gaps in employability models that were adequately discussed and treated comprehensively earlier in this book. Both variables were found to be critical in the employability discussion. The next section adequately discusses the four factors mentioned earlier, namely GS, DSS, WIL, and EI and how they contribute to graduate employability.

Generic skills (GS)

One of the critical dimensions that contribute to the employability of students is the possession of generic skills. Considering the various dynamics and complexities of the industry, including increased global competition, technological advancements, environmental factors, and construction projects sophistication, the need for skilled graduates cannot be overstated (Ahn *et al.*, 2012). According to Arain (2010), the industry of today is in dire need of graduates with a good foundation in construction principles and who can handle and oversee construction projects. They are further required to have ample industry experience as well as the ability to contribute meaningfully to design in improving the built environment (Russell *et al.*, 2007; Arain, 2010; Ahn *et al.*, 2012; Nowiński and Haddoud, 2019). Owing to the complexities of the construction industry, graduates are not only required to possess a sound academic background but also to possess non-academic skills and competencies to be able to thrive after graduation. These non-academic skills include personal values, communication skills, problem-solving skills, analytical abilities, interpersonal skills, teamwork skills, general initiative, critical thinking skills, soft skills, independence, decision-making skills, enterprise and adaptability skills, time management, work ethics, planning and organizing skills, numeracy skills, and technical skills (Archer and Davison, 2008; Farooqui and Ahmed, 2009; Arain, 2010; Wickramasinghe and Perera, 2010; Rawlins and Marasini, 2011; Ahn *et al.*, 2012; Ayarkwa *et al.*, 2012; Lievens and Sackett, 2012; Reid and Anderson, 2012; Jackson and Chapman, 2012; Samavedham and Ragupathi, 2012; Finch *et al.*, 2013; O'Leary, 2017; Nguyen, 2019). Other generic skills considered as important from extant literature include information and technology (ICT) skills, entrepreneurship skills, leadership skills, and ethical skills (Russell *et al.*, 2007; Archer and Davison, 2008; Tatum, 2010; Arain, 2010; Ahn *et al.*, 2012; Conrad and Newberry, 2012; Durrani and Tariq, 2012). Generic skills are also referred to as 'key skills', 'essential skills', 'core skills', 'soft skills', and 'transferrable skills'. However, the term 'generic skills' has been adopted in this book and has been hypothesized for the development of an employability improvement framework, as summarized in Table 7.1.

Table 7.1 Conceptual model latent constructs

Latent variable constructs	Measurement variables	Authors
Generic skills (GS)	Communication skills	Graham *et al.*(2009); Arain (2010); Jackson and Chapman (2012)
	Critical thinking skills	Kilgour and Koslow (2009); Wickramasinghe and Perera (2010); Conrad and Newberry (2012)
	Teamwork skills	Russell *et al.* (2007); Arain (2010); Jackson and Chapman (2012)
	Interpersonal skills	Arain (2010); Lievens and Sackett (2012)
	Leadership skills	Graham *et al.* (2009); Conrad and Newberry (2012)
	Adaptability skills	Arain (2010)
	Creativity/innovation skills	Wickramasinghe and Perera (2010); Jackson and Chapman (2012)
	Willingness to learn	Arain (2010)
	Organizing skills	Jackson and Chapman (2012)
	Management skills	Ahn *et al.* (2012)
	Information and Communication Technology (ICT) skills	Ahn *et al.* (2012); Zhou and Purushothaman (2019)
	Decision-making skills	Jackson and Chapman (2012)
	Analytical skills	Kilgour and Koslow (2009); Wickramasinghe and Perera (2010)
	Problem-solving skills	Jackson and Chapman (2012); Ahn *et al.* (2012); Reid and Anderson (2012)
	Entrepreneurship skills	Nowiński and Haddoud (2019)
	Networking skills	Wickramasinghe and Perera (2010)
Discipline-specific skills (DSS)	Knowledge of subject area	Jackson and Chapman (2012)
	Knowledge of overall curricula	Russell *et al.* (2007)
	Realizing strength and weaknesses	Arain (2010)
	Coordinating skills	Jackson and Chapman (2012)
	Administrative skills	Farooqui and Ahmed (2009)
	Lifelong learning	Casner-Lotto and Barrington (2006)
	Digital literacy	Arain (2010)
	Developing relationships with peers	Farooqui and Ahmed (2009)
	Self-confidence	Farooqui and Ahmed (2009); Arain (2010)

(Continued)

Table 7.1 (Continued)

Latent variable constructs	Measurement variables	Authors
	Exposure to other disciplines	Farooqui and Ahmed (2009)
	Hands-on experience	Jackson and Chapman (2012); Ahn et al. (2012)
	Technical skills	Jackson and Chapman (2012); Ahn et al. (2012)
Work-integrated learning (WIL)	Developed practical knowledge	Garavan and Murphy (2001); Lam and Ching (2007); Lowden et al. (2011)
	Understanding of job responsibilities	Mihail (2006); Lam and Ching (2007)
	Improved work attitude	Mihail (2006); Lowden et al. (2011)
	Developed interpersonal values	Mihail (2006); Lowden et al. (2011)
	Knowledge of résumé design	Lam and Ching (2007)
	Awareness of workplace culture	Garavan and Murphy (2001)
	Understanding of professional ethics	Mihail (2006)
	Developing professional identity	Wasonga and Murphy (2006)
	Knowledge of ongoing issues	Wasonga and Murphy (2006)
	Knowledge of safety procedures	Wasonga and Murphy (2006); Sattler et al. (2011)
	Exposure to multidisciplinary teams	Wasonga and Murphy (2006); Sattler et al. (2011)
	Knowledge of engineering designs	Sattler et al. (2011)
	Knowledge of quality control measures	Wasonga and Murphy (2006); Sattler et al. (2011)
	Site-supervision knowledge	Mihail (2006)
	Financial management skill	Omar et al. (2008)
	Networking with other interns on-site	Wasonga and Murphy (2006); Sattler et al. (2011)
	Exposure to career opportunities	Omar et al. (2008); Lowden et al. (2011)
	Enhanced learning	Sattler et al. (2011)
	Acquisition of professional skills	Mihail (2006); Lowden et al. (2011)
	Technical report writing	Wasonga and Murphy (2006); Sattler et al. (2011)
	Mentorship opportunities	Omar et al. (2008); Lowden et al. (2011)
Emotional intelligence (EI)	Perceiving and accurately expressing emotions	Goleman (1998); Blickle and Witzki (2008)
	Understanding and managing the emotions of others	Coetzee and Beukes (2010)
	Stability in making critical decisions	Stein and Book (2011); Jameson et al. (2016)

Latent variable constructs	Measurement variables	Authors
	Ability to motivate oneself	Stein and Book (2011); Jameson *et al.* (2016)
	Communicate with stable emotions	Coetzee and Beukes (2010)
	Ability to show compassion to others	Cherniss (2000); Lopes *et al.* (2006)
	Emotional reasoning	Coetzee and Beukes (2010)
	Ability to display a positive outlook to life	Cherniss (2000); Lopes *et al.* (2006)
	Flexibility in solving problems	Dacre Pool and Sewell (2007); Jameson *et al.* (2016)
	Building effective relationships	Stein and Book (2011); Jameson *et al.* (2016)
	Conflict management skills	Dacre Pool and Sewell (2007)
	Ability to manage stress	Dacre Pool and Sewell (2007)
	Ability to adapt to new ideas and surroundings	Cherniss (2000); Lopes *et al.* (2006); Jameson *et al.* (2016)
	Ability to display resilience in the face of adversity	Stein and Book (2011); Jameson *et al.* (2016)
University–industry collaboration (UIC)	Scholarship opportunities for students	Ramakrishnan and Yasin (2011)
	Job opportunities for graduates	Turhan and Akman (2013)
	Research grants from industry	Ankrah and Omar (2015)
	Inclusion of open-day events	Ramakrishnan and Yasin (2011); Ankrah and Omar (2015)
	Field trips to industry	Ankrah and Omar (2015); Haupt (2012)
	Industry mentoring for students	Ramakrishnan and Yasin (2011)
	In-service training	Ankrah and Omar (2015)
	Inviting guest speakers from industry	Marotta *et al.* (2007); Esham (2008)
	Networking activities with industry professionals	Marotta *et al.* (2007); Esham (2008)
	Entrepreneurial culture in universities	Haupt (2012)
	Publication opportunities with industry	Marotta *et al.* (2007)
	Construction project exercises in classrooms	Haupt (2012); Ankrah and Omar (2015)
	Patenting opportunities	Ramakrishnan and Yasin (2011); Ankrah and Omar (2015)

(Continued)

Table 7.1 (Continued)

Latent variable constructs	Measurement variables	Authors
	Training programme opportunities for students	Ramakrishnan and Yasin (2011); Ankrah and Omar (2015)
	Students' exposure to innovative ideas	Ankrah and Omar (2015)
	Involvement in construction projects	Marotta *et al.* (2007); Esham (2008); Haupt (2012)
Industry 4.0 (4IR) knowledge	Knowledge of the use of drones	Sundmaeker *et al.* (2010); Henning (2013); Kagermann *et al.* (2013); Hermann *et al.* (2016); Du Plessis (2017)
	Internet of Things (Use of Internet)	Sundmaeker *et al.* (2010); Henning (2013)
	Building information modelling (BIM)	Sundmaeker *et al.* (2010)
	Robotics knowledge	Hermann *et al.* (2016); Du Plessis (2017)
	Internet of services (ability to use software)	Kagermann *et al.* (2013); Hermann *et al.* (2016); Du Plessis (2017)
	Modelling-based tools (e.g. multivariate regression)	Du Plessis (2017)
	Advanced data analytics (knowledge of statistics and programming)	Sundmaeker *et al.* (2010); Henning (2013)
	Virtual reality (sophisticated headsets, LCD displays, etc.)	Du Plessis (2017)
	Augmented reality (use of smartphones, video games, etc.)	Kagermann *et al.* (2013); Hermann *et al.* (2016); Du Plessis (2017)
	Mixed reality (understanding holograms, 3D modelling, etc.)	Sundmaeker *et al.* (2010); Henning (2013)
	Knowledge of cybersecurity	Kagermann *et al.* (2013); Hermann *et al.* (2016); Du Plessis (2017)
	Knowledge of artificial intelligence	Kagermann *et al.* (2013); Hermann *et al.* (2016); Du Plessis (2017)
	Knowledge of cryptocurrency blockchain/bitcoin	Kagermann *et al.* (2013); Hermann *et al.* (2016); Du Plessis (2017)
	Engineering design (computer-aided design)	Hermann *et al.* (2016); Du Plessis (2017)
	Electronics technology (ability to apply basic electronic principles)	Hermann *et al.* (2016); Du Plessis (2017)
	Self-service use of E-library	Hermann *et al.* (2016); Du Plessis (2017)
	Cloud computing	

Latent variable constructs	Measurement variables	Authors
	Knowledge of administrative tools (e.g. Microsoft Suites, Primavera)	Hermann *et al.* (2016); Du Plessis (2017)
	Information technology and systems control	Hermann *et al.* (2016); Du Plessis (2017)

Discipline-specific skills (DSS)

This dimension refers to a set of knowledge and understanding which is specific to an academic discipline or career profession. According to Bridgstock (2009), they are those sets of skills, knowledge, and competencies that are embedded in a specific discipline to address specific job-related requirements. These skills usually originate in subject matter areas, specific domains, and disciplines. For instance, a civil engineering graduate should have the ability to apply principles to the engineering practice to design and supervise construction projects after graduation. Also, a graduate in quantity surveying (QS) should possess the ability to conduct feasibility studies to estimate materials, time, and labour costs. A graduate QS should also be able to prepare, negotiate, and analyse costs for tenders and contracts. Therefore, for built environment graduates to obtain employment from the industry successfully after graduation, they are required to possess a foundational knowledge within that field. Students can obtain discipline-specific skills or knowledge during lectures, simulations, tutorials, and work-integrated learning via face-to-face approaches or online means. Some of the knowledge and skills acquired by studying an academic discipline include knowledge of subject area and curricula, critical thinking, pedagogical skills, coordination skills, communication skills, administrative skills, management skills, lifelong learning, decision-making skills, digital literacy, self-confidence, exposure to other disciplines, problem-solving skills, collaborative skills, hands-on experience, and technical skills as confirmed by Dacre Pool and Sewell (2007), Rae (2007), Bridgstock (2009), Coetzee and Beukes (2010), Hutchinson (2013), Watts (2013), Nagarajan and Edwards (2015), and Lamanauskas and Augienė (2017).

Work-integrated learning (WIL)

The importance of work experience for built environment graduates cannot be quantified. By engaging in work activities, students are presented with enhanced knowledge about their career choices and related occupations. Hence, these opportunities help students ascertain the strengths, weaknesses, and expectations of their chosen fields and how they can explore other possible options. According to Wasonga and Murphy (2006) and Sattler *et al.* (2011), work activities help students to apply information garnered from lecture-room activities which further improve their learning and thought processes as they commence their professional careers. Omar *et al.* (2008) opine that students who undergo work programmes are more attractive to potential employers after graduation. Students who experience work activities are also enriched with increased practical knowledge, social skills, and marketability after graduation as well as increased understanding of the construction industry needs and expectations (Sattler *et al.*, 2011).

Similarly, Gault *et al.* (2000) and Wasonga and Murphy (2006) suggest that work programmes can lead to improved academic performance among students because students' interest and enthusiasm in the classroom are enhanced when work activities are integrated. Likewise, it helps to build a stronger résumé for job applications (Divine *et al.*, 2007; Lowden *et al.*, 2011); it illuminates students' career choices upon graduation (Lowden *et al.*, 2011) and helps candidates express confidence during job interviews (Pillai *et al.*, 2012) and display self-reliance and maturity (Gill and Lashine, 2003; Mihail, 2006; Pillai *et al.*, 2012). Likewise, McLennan and Keating (2008), Orrell (2004), and Cooper *et al.* (2010) add that it helps to build interpersonal relations, teamwork skills, career development, adaptability skills, and building network contacts, amongst others.

Some of these knowledge and skills acquired by undergoing work experience (work-integrated learning) include understanding job responsibilities, awareness of industry, improved work attitude, awareness of workplace culture, time management, self-confidence, problem-solving ability, working in multidisciplinary teams, site-supervision knowledge, financial management skills, interpersonal relationships with industry professionals, acquisition of professional skills, technical report writing, and mentorship opportunities. This set of skills confirms the works of Gault *et al.* (2000), Gill and Lashine (2003), Mihail (2006), Wasonga and Murphy (2006), Divine *et al.* (2007), McLennan and Keating (2008), Omar *et al.* (2008), Cooper *et al.* (2010), Sattler *et al.* (2011), Lowden *et al.* (2011), and Pillai *et al.* (2012). For this book, the term 'work-integrated learning' (WIL) has been adopted. It is an umbrella term used to describe a more situated, participative, and 'real-world' oriented programme that integrates conventional learning and workplace concerns. There is a wide range of approaches and pedagogical forms through which WIL enhances student learning and employability. They include action-learning, cooperative education, apprenticeships, experiential learning, inter-professional learning, inquiry learning, practicum placements, project-based learning, practice-based learning, scenario learning, work experience, workplace learning, internship, vocational learning, and service-learning, amongst others (Sattler, 2011). Though the terminologies used to describe the programmes and practices may vary, they all have a common theme, which highlights the significance of work activities in complementing theoretical knowledge obtained during formal lectures. Students who undergo WIL tend to experience the following: improved academic performance, increased motivation to learn, personal skills and competencies, enhancement of interdisciplinary thinking, development of work ethics, and increased technical know-how.

Emotional intelligence (EI)

Another key dimension for graduate employability is emotional intelligence. Goleman (1998) and Coetzee and Beukes (2010) describe EI as the capacity to recognize one's feelings and that of others. Stein and Book (2011) define it as 'a set of skills that enable us to make our way in a complex world – the personal, social and survival aspects of overall intelligence, the elusive common sense and sensitivity that are essential to effective daily functioning'. Similarly, Cherniss (2000), Lopes *et al.* (2006), and Dacre Pool and Sewell (2007) suggest that individuals with high levels of EI tend to motivate others and themselves which can lead to successful careers. This was resonated by Coetzee and Beukes (2010), who assert that an increased level of EI leads to increased confidence in exhibiting skills at the workplace. In facing the present-day industry and environment demands, students must possess EI and social skills. Bar-On (1997) highlights it as an encompassing adaptability that suggests that individuals with high EI can possibly manage people with a lower level. Blickle and Witzki (2008) and Jameson *et al.* (2016) also suggest that exhibiting optimism, self-confidence, self-management, self-awareness, social awareness, and internal locus of control are necessary to adapt successfully to varying work conditions. This firmly relates to self-efficacy which is based on an individual's belief in oneself to cope and thrive in the changing world of work. Therefore, emotionally intelligent individuals are more likely to be employable and successful in their chosen careers (Fugate and Ashforth, 2004). The mixed model of EI consists of both ability- and trait-related constructs, including interpersonal skills, intrapersonal skills, stress management skills, adaptability skills, and a generally positive mood (Bar-On, 1997). Furthermore, individuals who exhibit high levels of emotional intelligence have been found to perceive and accurately express emotion, make decisions effectively, communicate in a stabilized manner, manage relationships, show compassion for others, solve problems flexibly, build effective relationships, adapt to new ideas and surroundings, manage stress, and display resilience in the face of adversity. This set of skills confirms the works of Goleman (1998); Cherniss (2000), Fugate and Ashforth (2004), Lopes *et al.* (2006), Dacre Pool and Sewell (2007), Blickle and Witzki (2008), Coetzee and Beukes (2010), Stein and Book (2011), and Jameson *et al.* (2016).

Framework specification and justification

Despite the debates and reviews surrounding the concept of graduate employability, there has been little mention of an employability skills framework specifically for the built environment domain in the 4IR era, a gap this book intends to address. The framework for this book synthesizes the seminal contributions of Law and Watts's (1997) DOTS model, the heuristic model of employability of Fugate and Ashforth (2004), Van Dam's (2004) employability orientation process model, McQuaid and Lindsay's (2005) employability framework, Yorke and Knight's (2006) USEM model, Dacre Pool and Sewell's (2007) CareerEDGE model, Bridgstock's (2009) conceptual model of graduate attributes for employ-ability, Coetzee's (2008) psychological career resources model, and Bezuiden-hout's (2011) graduate employability model.

However, for the sake of this book, the 'career development learning' (CDL) dimension and the 'subject knowledge' (SK) dimension will be regarded as 'discipline-specific skills'. The reason for this is that CDL involves the activities that help to develop a student's career in relation to skills and knowledge which can be directly or indirectly obtained from subject or discipline modules. Simi-larly, information and ideas obtained during discipline-specific lectures can help students to become more self-aware, understand their interests with regard to life planning, develop skills/competencies, understand their strengths and weaknesses and areas which require the need for improvement, and even improve their career attitudes. However, neither the external factors nor personal circumstances were considered. This is because higher education has no control over the influences of these dimensions on the employability of graduates.

Most importantly, some of these frameworks have become dated because the concept of employability is constantly evolving as present-day graduates need to possess more proficiencies to function effectively in this dynamic era of work, coupled with the recent wave of the 4IR. Of the nine models mentioned previ-ously, the CareerEDGE model seems to encapsulate several dimensions that apply to the wider and contextual study of the concept of employability. These con-structs include the possession of degree subject knowledge, understanding, and skills (discipline-specific skills), career development learning, generic skills, work experience, emotional intelligence, and the 3S constructs (self-efficacy/self-confidence/self-esteem). For graduates to be adequately prepared to fit into the world of work, they need to possess these elements, according to Dacre Pool and Sewell (2007). Therefore, the conceptual framework for this study is based on the critical components of the 'CareerEDGE' model, namely generic skills (GS), discipline-specific skills (DSS), work-integrated learning (WIL), and emotional intelligence (EI). The additional two constructs (the gaps) are university–industry collaboration (UIC) and Industry 4.0 knowledge (4IR). It is expected that these variables will result in improved graduate employability (IGE). In this regard, IGE was treated as a criterion variable, and hence a dependent variable.

Based on the main dimensions and constructs peculiar to the nine existing models of employability reviewed (Table 7.2), the proposed employability frame-work examines the relationship of the generic skills, discipline-specific skills, work-integrated learning, and emotional intelligence (essential variables mea-sured from previous studies), with the inclusion of university–industry collabora-tion and 4IR knowledge (exogenous variables) and their roles in ensuring improved graduate employability (endogenous variable). These will, in turn, result in industry-ready graduates with improved problem-solving abilities and who can meet the needs of the South African construction industry, notwith-standing the advent of the 4IR.

Another objective of this book is to forecast the relative predictive power of these main dimensions (exogenous variables) for improved graduate employability

Framework underpinning employability   177

Table 7.2 List of various dimensions peculiar to reviewed models

Authors/year	Generic skills	Career development learning	Subject knowledge	Work experience	Emotional intelligence	External factors	Personal circumstances
1 Law and Watts (1997)	X	X	X	X			
2 Fugate and Ashforth (2004)	X	X				X	X
3 Van Dam (2004)	X	X		X			
4 McQuaid and Lindsay (2005)	X					X	X
5 Yorke and Knight (2006)	X	X	X		X		
6 Dacre Pool and Sewell (2007)	X	X	X	X	X		
7 Bridgstock (2009)	X	X	X	X	X		
8 Coetzee (2008)	X	X	X		X		
9 Bezuidenhout (2011)	X	X	X		X		

to test (determine) whether graduate employability depends on the constructs of these variables, taking into account the effects of the benefits obtained when university and industry collaborate and 4IR knowledge are present. The conceptual framework theorizes that improved graduate employability is established by the existing relationship between the main (exogenous) variables and its underlying factors by which both the objective and subjective measures are linked. These variables, key factors, and constructs were identified from reviewed models and existing literature. Hence, they are considered to be the critical determinants of improved graduate employability for built environment graduates.

Structural components of the framework

The present conceptual framework theorizes that improved graduate employability (IGE) for built environment graduates is derived from the possession of generic skills (GS), discipline-specific skills (DSS), work-integrated learning (WIL), emotional intelligence (EI), university–industry collaboration (UIC) outcomes, and Industry 4.0 knowledge (4IR). The employability framework to be tested in the postulated hypothesis has never been tested and is a multidimensional structure composed of GS, DSS, WIL, EI, UIC, and 4IR. This postulated framework is shown in Figure 7.1. The conceptual framework is based on the critical components of the 'CareerEDGE' model (Dacre Pool and Sewell, 2007). In this conceptualized framework, improved graduate employability of built environment graduates is related to the evaluation of several variables. It is difficult to discuss the generic skills without referring to the variables of discipline-specific skills, work experience factors, and even emotional intelligence factors. It remains to be seen which of these exogenous variables is the most relevant because the responses of individuals may vary under different circumstances. Also, different job descriptions require a different skill set. Therefore, employers in the construction industry may vary in their opinions regarding which of these skills and competencies are most relevant. One thing is certain: the advent of the 4IR will ultimately prompt employers to search for multi-skilled graduates. This means graduates will have to possess a wide range of skills (generic and digital) and knowledge to be deemed employable.

Outcomes of the employability model

As shown in Figure 7.1, the employability framework posits that improved IGE for built environment graduates is derived from the possession of GS, DSS, WIL, EI, UIC outcomes, and 4IR. One of the outcomes of the IGE is industry-readiness. As earlier argued, the possession of both academic and non-academic skills is pivotal in ensuring the industry-readiness of graduates after graduation. Industry-ready graduates are usually equipped with the right attributes, skills, and mindset to make meaningful contributions to the world of work when they are called upon. Likewise, the possession of both academic and non-academic skills is pivotal in preparing students to perform at the top level in their various careers after

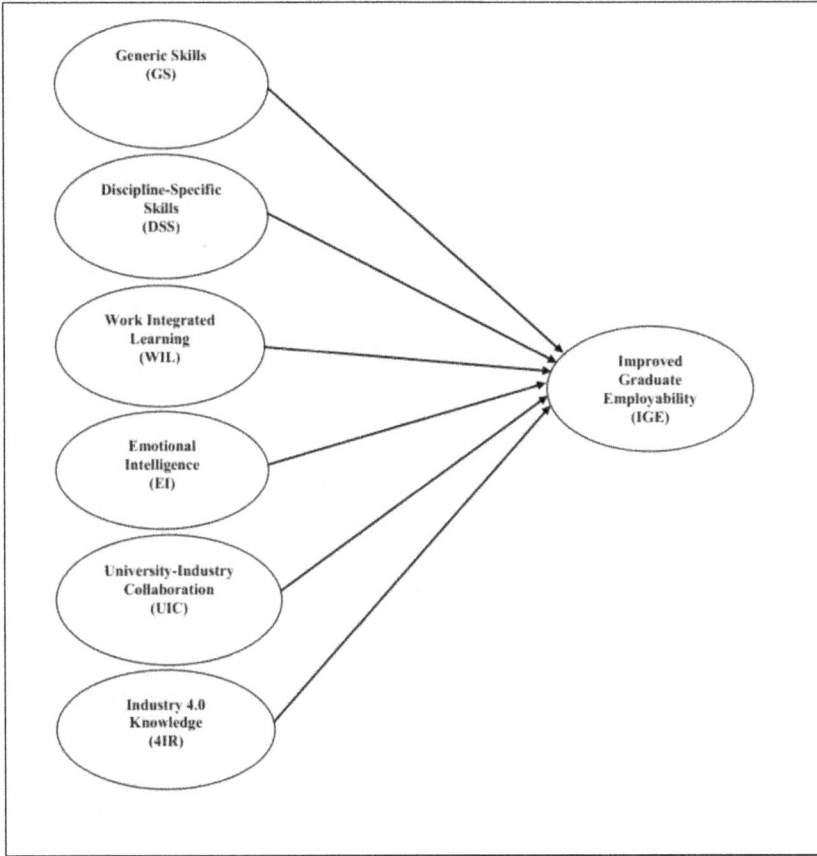

Figure 7.1 An employability improvement framework (EIF)

graduation. Graduates who are industry-ready possess the tendency to become reliable in carrying out their tasks effectively and hence are responsible. As an outcome of IGE, it is expected that the future workforce will possess a sense of environmental, social, ethical, and economic responsibility towards the society they intend to serve. Moreover, it is expected that emotional intelligent graduates will establish better collaboration with peers and professionals as they can perceive and express emotions accurately as well as understand and manage the emotions of others. Emotional intelligent graduates can also communicate with stable emotions and can build healthy relationships with people. Thus, they can become future leaders who can manage people, display confidence in addressing issues, and work effectively under pressure (Blickle and Witzki, 2008; Jameson *et al.*, 2016).

Additionally, graduates who possess an array of generic skills, discipline-specific skills, and emotional intelligence are technically sound and can adapt easily to changing work conditions which are outcomes of IGE. Owing to the various complexities of the construction industry, present-day employers are seeking graduates who can adapt to varying circumstances and can seamlessly understand new ideas and concepts seamlessly. Graduates with such abilities can deal with unfamiliar domains as well as thrive in a culture of change. Hence, another outcome of IGE is the increased flexibility of graduates. Flexible graduates can improve the way they think and work to meet the increasing expectations of industry employers. Moreover, the possession of an array of skills and knowledge within the models is expected to boost the inventiveness and critical thinking skills of graduates.

In an increasingly global marketplace, present-day graduates will be expected to have a global perspective and be ready to work with various teams in achieving timely results (Archer and Davison, 2008). Today's ethical issues are assuming global proportions, and graduates must possess a strong ethical foundation needed to deal with issues involving the equitable distribution of resources, by-products of design, sustainable development, and environmental conservation, amongst others (Arain, 2010; Mat and Zabidi, 2010; Ma and Sun, 2013). They also need to be conversant with the legal aspects of engineering and its social impact as well as a proper understanding of business practices and entrepreneurship ideas (Jackson and Chapman, 2012). The possession of these skills and knowledge will enhance their employment prospects as well as the understanding of the socio-economic implications of their chosen careers. Ultimately, the presence of these key skills and competencies among graduates can lead to increased efficiency and productivity within the industry. Furthermore, graduates with improved knowledge of 4IR components will be highly sought after by employers. As one of the variables which contribute to IGE, it is necessary that graduates possess a wide range of 4IR skills and knowledge to be deemed employable in the near future.

Therefore, both education and training sectors are presented with the daunting task of aligning themselves with the construction industry to be conversant with the ever-changing skill demands of employers (UIC collaboration which is highlighted in Figure 7.1). Hence, this book proposes that IGE is a function of required employability attributes and skills (generic, discipline-specific, work experience, emotional intelligence, UIC collaboration, and 4IR knowledge). Ultimately, this leads to a greater pool of skilled graduates from which employers can choose.

Measurement components of the model

The measurement component of the hypothesized model comprises the following improved graduate employability factors: generic skills (GS) with 16 measurement variables; discipline-specific skills (DSS) with 12 measurement variables; work-integrated learning (WIL) with 21 measurement variables; emotional intelligence (EI) with 14 measurement variables; university–industry collaboration (UIC) outcomes with 16 measurement variables, and Industry 4.0 (4IR) knowledge with 19 measurement variables. From this framework, the success of built environment graduates can be measured and is guaranteed if they possess these various skills and knowledge after graduation.

Summary

This chapter discussed the conceptual framework underpinning this study. The hypothesized employability improvement framework was also presented in this chapter based on a rigorous review of nine previous employability models, which was presented earlier in this book. Generic skills (GS), discipline-specific skills (DSS), work-integrated learning (WIL), and emotional intelligence (EI) were adopted from the reviewed models and as main constructs. The theorized model proposes that IGE is a multidimensional structure composed of six latent variables: generic skills (GS), discipline-specific skills (DSS), work-integrated learning (WIL), emotional intelligence (EI), university–industry collaboration (UIC), and Industry 4.0 (4IR) knowledge. These six latent variables are also regarded as the critical success factors for improved graduate employability. The next chapter focuses on the validation of these factors by adopting a Delphi approach using South Africa as a case study.

References

Ahn, Y.H., Annie, R.P. and Kwon, H. 2012. Key competencies for US construction graduates: Industry perspective. *Journal of Professional Issues in Engineering Education and Practice, 138*(2), pp. 123–130. doi: 10.1061/(ASCE)EI.1943-55410000089.

Ankrah, S. and Omar, A.T. 2015. Universities-industry collaboration: A systematic review. *Scandinavian Journal of Management, 31*(3), pp. 387–408.

Arain, F.M. 2010. Identifying competencies for baccalaureate level construction education: Enhancing employability of young professionals in the construction industry. In *Proceedings of the Construction Research Congress 2010: Innovation for Reshaping Construction Practice*, Construction Research Congress 2010, American Society of Civil Engineers (ASCE), Banff, AB, p. 194.

Archer, W. and Davison, J. 2008. *Graduate employability. What employers want?* London: The Council for Industry and Higher Education.

Ayarkwa, J., Adinyira, E. and Osei-Asibey, D. 2012. Industrial training of construction students: Perceptions of training organizations in Ghana. *Education Training, 54*(2/3), pp. 234–249. http://dx.doi.org/10.1108/00400911211210323.

Bar-On, R. 1997. *The Bar-On emotional quotient inventory (EQ-i): A test of emotional intelligence.* Toronto: Multi-Health Systems.

Bezuidenhout, M. 2011. *The development and evaluation of a measure of graduate employability in the context of the new world of work* (Doctoral dissertation, University of Pretoria, Pretoria).

Blickle, G. and Witzki, A. 2008. New psychological contracts in the world of work: Economic citizens or victims of the market? *Society and Business Review, 3*(2), pp. 149–161.

Bridgstock, R. 2009. The graduate attributes we've overlooked: Enhancing graduate employability through career management skills. *Higher Education Research and Development, 28*(1), pp. 31–44. doi: 10.1080/07294360802444347.

Casner-Lotto, J. and Barrington, L. 2006. *Are they really ready to work? Employers' perspectives on the basic knowledge and applied skills of new entrants to the 21st century US workforce.* Washington, DC: Partnership for 21st Century Skills.

Cherniss, C. 2000. Emotional intelligence: What it is and why it matters. *Paper presented at the annual meeting of the Society for Industrial and Organizational Psychology,* New Orleans, USA.

Coetzee, M. 2008. Psychological career resources of working adults: A South African survey. *South African Journal of Industrial Psychology,* 34(2), pp. 10–20.

Coetzee, M. and Beukes, C.J. 2010. Employability, emotional intelligence and career preparation support satisfaction among adolescents in the school-to-work transition phase. *Journal of Psychology in Africa,* 20(3), pp. 439–446.

Conrad, D. and Newberry, R. 2012. Identification and instruction of important business communication skills for graduate business education. *Journal of Education for Business,* 87(2), pp. 112–120.

Cooper, L., Orrell, J. and Bowden, M. 2010. *Work integrated learning: A guide to effective practice.* New York: Routledge.

Dacre Pool, L. and Sewell, P. 2007. The key to employability: Developing a practical model of graduate employability. *Education Training,* 49(4), pp. 277–289.

Divine, R.L., Linrud, J.K., Miller, R.H. and Wilson, J.H. 2007. Required internship programs in marketing: Benefits, challenges and determinants of fit. *Marketing Education Review,* 17(2), pp. 45–52. http://dx.doi.org/10.1080/10528008.2007.11489003.

Du Plessis, C.J. 2017. *A framework for implementing Industry 4.0 in learning factories* (Doctoral dissertation, Stellenbosch University, Stellenbosch).

Durrani, N. and Tariq, V.N. 2012. The role of numeracy skills in graduate employability. *Education Training,* 54(5), pp. 419–434. http://dx.doi.org/10.1108/00400911211244704.

Esham, M. 2008. Strategies to develop university-industry linkages in Sri Lanka. *National Education Commission Sri Lanka Study Series No. 4 (2007).*

Farooqui, R.U. and Ahmed, S.M. 2009. Key skills for graduating construction management students: A comparative study of industry and academic perspectives. *Proceedings of the Construction Research Congress 2009: Building a Sustainable Future,* American Society of Civil Engineers (ASCE), Washington, DC, p. 1439.

Finch, D.J., Hamilton, L.K., Baldwin, R. and Zehner, M. 2013. An exploratory study of factors affecting undergraduate employability. *Education Training,* 55(7), pp. 681–704. http://dx.doi.org/10.1108/ET-07-2012-0077.

Fugate, M. and Ashforth, B. 2004. Employability: The construct, its dimensions, and applications. *Proceedings of the Annual Meeting of the Academy of Management,* OB: J1-J6, Seattle.

Garavan, T.N. and Murphy, C. 2001. The co-operative education process and organisational socialisation: A qualitative study of student perceptions of its effectiveness. *Education + Training,* 43(6), pp. 281–302.

Gault, J., Redington, J. and Schlager, T. 2000. Undergraduate business internships and career success: Are they related? *Journal of Marketing Education,* 22(1), pp. 45–53.

Gill, A. and Lashine, S. 2003. Business education: A strategic market-oriented focus. *International Journal of Educational Management,* 175, pp. 188–194. http://dx.doi.org/10.1108/09513540310484904.

Goleman, D. 1998. *Working with emotional intelligence.* New York: Bantam.

Graham, R., Crawley, E. and Mendelsohn, B.R. 2009. Engineering leadership education: A snapshot review of international good practice. In *White paper sponsored by the Bernard*

Gordon-MIT Engineering Leadership Program (p. 41). Cambridge, MA. Retrieved from: http://web.mit.edu/gordonelp/elewhitepaper.pdf [Accessed 30 November 2011].

Harvey, L. 2002. *Employability and diversity.* Retrieved from: www2.wlv.ac.uk/webteam/ confs/socdiv/sdd-harvey-0602.

Haupt, T. 2012. Student attitudes towards cooperative construction education experiences. *Construction Economics and Building, 3*(1), pp. 31–42.

Henning, K. 2013. *Recommendations for implementing the strategic initiative industrie 4.0.*

Hermann, M., Pentek, T. and Otto, B. 2016, January. Design principles for industrie 4.0 scenarios. *Proceedings of the System Sciences (HICSS), 2016 49th Hawaii International Conference*, IEEE, Piscataway, NJ, pp. 3928–3937.

Hillage, J. and Pollard, E. 1998 Employability: Developing a framework for policy analysis. *Research Brief 85.* London: Department for Education and Employment.

Hutchinson, J. 2013. *School organisation and STEM career-related learning.* Derby: International Centre for Guidance Studies. Retrieved from: http://derby.openrepository.com/ derby/bitstream/10545/303288/8/STEM%20Leaders%20Report%202013%20 %28High%20res%29.pdf.

Jackson, D. and Chapman, E. 2012. Non-technical competencies in undergraduate business degree programs: Australian and UK perspectives. *Studies in Higher Education, 37*(5), pp. 541–567. doi: 10.1080/03075079.2010.527935.

Jameson, A., Carthy, A., McGuinness, C. and McSweeney, F. 2016. Emotional intelligence and graduates-employers' perspectives. *Procedia – Social and Behavioural Sciences, 228*, pp. 515–522.

Kagermann, H., Wahlster, W. and Helbig, J. 2013. *Recommendations for implementing the strategic initiative Industrie 4.0.* Final report of the Industrie 4.0 working group.

Kilgour, M. and Koslow, S. 2009. Why and how do creative thinking techniques work? Trading off originality and appropriateness to make more creative advertising. *Journal of the Academy of Marketing Science, 37*(3), pp. 298–309.

Lam, T. and Ching, L. 2007. An exploratory study of an internship program: The case of Hong Kong students. *International Journal of Hospitality Management, 26*(2), pp. 336–351.

Lamanauskas, V. and Augienė, D. 2017. Career education in comprehensive schools of Lithuania: Career professionals' position. In A. Klim-Klimaszewska, M. Podhajecka and A. Fijalkowska-Mroczek (Eds.), *Orientacje i przedsiewziecia w edukacjiprzedszkolnej i szkolnej /Monografia/* (pp. 216–236). Siedlce: Akka.

Law, W. and Watts, A.G. 1977. *Schools, careers and community.* London: Church Information Office.

Lievens, F. and Sackett, P.R. 2012. The validity of interpersonal skills assessment via situational judgment tests for predicting academic success and job performance. *Journal of Applied Psychology, 97*(2), p. 460.

Lopes, P.N., Grewal, D., Kadis, J., Gall, M. and Salovey, P. 2006. Evidence that emotional intelligence is related to job performance and affect and attitudes at work. *Psichothemia, 18*(1), pp. 132–138.

Lowden, K., Hall, S., Ellio, D.D. and Lewin, J. 2011. *Employers' perceptions of the employability skills of new graduates.* London: Edge Foundation.

Ma, X. and Sun, Y. 2013. Research on employment ability evaluation of graduates. *Proceedings of the 19th International Conference on Industrial Engineering and Engineering Management*, Springer, p. 1019.

Marotta, D., Mark, M., Blom, A. and Thorn, K. 2007. *Human capital and university-industry linkages' role in fostering firm innovation: An empirical study of Chile and Colombia.* Retrieved from: https://openknowledeg.worldbank.org/bitstream/handle.

Mat, N.H.N. and Zabidi, Z.N. 2010. Professionalism in practices: A preliminary study on Malaysian public universities. *International Journal of Business and Management*, 5(8), pp. 138.

Maxwell, J.A. 2005. Conceptual framework: What do you think is going on? *Qualitative Research Design: An Interactive Approach*, *41*, pp. 33–63.

McLennan, B. and Keating, S. 2008, June. Work-integrated learning (WIL) in Australian universities: The challenges of mainstreaming WIL. *Proceedings of the ALTC NAGCAS National Symposium*, Melbourne, Australia, pp. 2–14.

McQuaid, R.W. and Lindsay, C. 2005. The concept of employability. *Urban Studies*, *42*(2), pp. 197–219.

Mihail, D.M. 2006. Internships at Greek universities: An exploratory study. *Journal of Workplace Learning*, *18*(1), pp. 28–41. doi: 10.1108/13665620610641292.

Miles, M.B. and Huberman, A.M. 1994. *Qualitative data analysis: An expanded sourcebook*. Thousand Oaks, CA: Sage.

Nagarajan, S. and Edwards, J. 2015. The role of universities, employers, graduates and professional associations in the development of professional skills of new graduates. *Journal of Perspectives in Applied Academic Practice*, *3*(2).

Nguyen, P. 2019. *Enhancing the employability of graduate students with transversal skills* (Bachelor's thesis, Lahti University of Applied Sciences, Lahti).

Nowiński, W. and Haddoud, M.Y. 2019. The role of inspiring role models in enhancing entrepreneurial intention. *Journal of Business Research*, 96, pp. 183–193.

O'Leary, S. 2017. Graduates' experiences of, and attitudes towards, the inclusion of employability-related support in undergraduate degree programmes; Trends and variations by subject discipline and gender. *Journal of Education and Work*, 30(1), pp. 84–105.

Omar, M.Z., Rahman, M.N.A., Kofli, N.T., Mat, K., Darus, M.Z. and Osman, S.A. 2008. Assessment of engineering students' perception after industrial training placement. *Proceedings of 4th WSEAS/IASME International Conference on Educational Technologies (EDUTE'08)*, Corfu, Greece.

Orrell, J. 2004. Work-integrated learning programmes: Management and educational quality. *Proceedings of the Australian Universities Quality Forum 2004*, 1–5, AUQA Occasional Publication, Australia.

Pillai, S., Khan, M.H., Ibrahim, I.S. and Raphael, S. 2012. Enhancing employability through industrial training in the Malaysian context. *Higher Education*, 63(2), pp. 187–204. doi: 10.1007/s10734-011-9430-2.

Rae, D. 2007. Connecting enterprise and graduate employability: Challenges to the higher education culture and curriculum? *Education+ Training*, 49(8/9), pp. 605–619.

Ramakrishnan, K. and Yasin, N.M. 2011. Higher learning institution-industry collaboration: A necessity to improve teaching and learning process. *Proceedings of the Computer Science and Education (ICCSE) 2011 6th International Conference on IEEE*, Singapore, p. 1445.

Rawlins, J. and Marasini, R. 2011. Are the construction graduates on CIOB accredited degree courses meeting the skills required by the industry? *Management*, *167*, pp. 174.

Reid, J.R. and Anderson, P.R. 2012. Critical thinking in the business classroom. *Journal of Education for Business*, 87(1), pp. 52–59.

Russell, J.S., Hanna, A., Bank, L.C. and Shapira, A. 2007. Education in construction engineering and management built on tradition: Blueprint for tomorrow. *Journal of Construction Engineering and Management*, *133*(9), pp. 661–668. doi: 10.1061/(ASCE)0733-9364(2007)133:9(661).

Samavedham, L. and Ragupathi, K. 2012. Facilitating 21st century skills in engineering students. *Journal of Engineering Education Transformations*, 25(4–1), pp. 37–49.

Sattler, P., Wiggers, R.D. and Arnold, C. 2011. Combining workplace training with post-secondary education: The spectrum of work-integrated earning (WIL) opportunities from apprenticeship to experiential learning. *Canadian Apprenticeship Journal*, 5. Retrieved from: www.adapt.it/fareapprendistato/docs/combining_workplace_training.pdf.

Stein, S.J. and Book, H.E. 2011. *The EQ edge: Emotional intelligence and your success*. Hoboken, NJ: John Wiley & Sons.

Sundmaeker, H., Guillemin, P., Friess, P. and Woelfflé, S. 2010. Vision and challenges for realising the Internet of Things. *Cluster of European Research Projects on the Internet of Things, European Commission*, 3(3), pp. 34–36.

Tatum, C. 2010. Construction engineering education: Need, content, learning approaches. In *Proceedings of the Construction Research Congress 2010: Innovation for Reshaping Construction Practice*, Construction Research Congress 2010, May 8–10, 2010, American Society of Civil Engineers (ASCE), Banff, AB, p. 183.

Turhan, C. and Akman, I. 2013. Employability of IT graduates from the industry's perspective: A case study in Turkey. *Asia Pacific Education Review*, 14(4), pp. 523–536. doi: 10.1007/s12564013-9278-5.

Van Dam, K. 2004. Antecedents and consequences of employability orientation. *European Journal of Work and Organizational Psychology*, 13(1), pp. 29–51.

Wasonga, T.A. and Murphy, J.F. 2006. Learning from tacit knowledge: The impact of the internship. *International Journal of Educational Management*, 20(2), pp. 153–163.

Watts, A.G. 2013. Career guidance and orientation. In the United Nations Educational, Scientific and Cultural Organization (UNESCO). *Revisiting Global Trends in TVET: Reflections on Theory and Practice*. Germany: UN Campus. Retrieved from: www.unevoc.unesco.org/fileadmin/up/2013_epub_revisiting_global_trends_in_tvet_book.pdf.

Wickramasinghe, V. and Perera, L. 2010. Graduates', university lecturers' and employers' perceptions towards employability skills. *Education+ Training*, 52(3), pp. 226–244.

Yorke, M. and Knight, P. 2006. *Embedding employability into the curriculum*. York: Higher Education Academy.

Zhou, C. and Purushothaman, A. 2019. Developing creativity and learning design by Information and Communication Technology (ICT) in developing contexts. In *Advanced methodologies and technologies in artificial intelligence, computer simulation, and human-computer interaction* (pp. 499–511). Hershey, PA: IGI Global.

8 Validation of the employability framework constructs

Overview of the Delphi method

The Delphi technique came into prominence in the 1950s at the beginning of the Cold War. At the time, it was adopted as a tool to forecast the impact of technological advancements on warfare as well as solve complex problems at the Rand Corporation (Buckley, 1995). Factually, the term 'Delphi' originates from the Oracle of Delphi – an ancient Greek temple where the oracle was located. According to historians from Greek mythology, the Oracle of Delphi was regularly consulted to forecast and predict the future to ensure accurate and timely planning before launching an attack or executing a tactical move. This method allowed subject-matter experts at the Rand Corporation to provide their opinions on probabilities, intensities, and likelihoods of possible enemy incursions and attacks. This predictive process was repeated severally until a consensus was achieved. Based on the historical antecedence, the Delphi technique is a qualitative methodology aimed at reaching consensus on a specific subject matter through a multi-round survey involving a group of experts in their various fields of influence. During the Delphi process, there are two or more 'rounds' in which the second round or later rounds are referred to as 'feedbacks'. Thus, experts can modify their responses based on the opinions of other experts. Hence, the Delphi method is usually a controlled group process designed to obtain the opinions of experts on subject matter where information is not holistically available or absolutely accurate (Häder and Häder, 1995). During the Delphi method, the knowledgeable panel (experts) responds to the questions and reverts the answers to the researcher (coordinator of the study) who analyses (processes) the responses and determines the central and extreme tendencies required for the study (Grisham, 2008). The experts, who remain anonymous throughout the process, have the opportunity to resubmit their opinions as many times (rounds) as possible until the researcher achieves a reliable consensus on the questions asked. Each round presents its own feedback on the statistical group judgement based on median, percentages, and the interquartile range. In the case of extreme opinions, there is room for arguments and counter-arguments that can be considered by the researcher. The Delphi method is often used to make predictions, inquiries, and futuristic projections. In addition, it often forms part of the policymaking approach which includes the nominal group technique (NGT) and interacting group method (IGM). This Delphi function has amplified its popularity and adoption across several scientific and academic disciplines as a study instrument and systematic thinking tool. The method has also gained valuable recognition from the expert community for producing creditable and credible outcomes.

Considering that the outcome of this book might trigger other aspects for future research and that it is aimed at engaging geographically dispersed experts from around South Africa, the Delphi method is rightly considered as a research approach and method. Moreover, employability studies in South Africa or other developing countries have rarely considered the Delphi technique. Because the nature of the research questions was difficult to measure except through experimental methods, the Delphi method was considered most appropriate. Moreover, for this current research study, the adoption of an experimental survey was not feasible. The rigorousness of the Delphi method (initial feedback from experts and subsequent distribution for further review) also makes this technique a robust approach and more appropriate than ordinary survey research (Stitt-Gohdes and Crews, 2004). According to Loo (2002), the Delphi method is most appropriate for policymakers who will steer the future direction of certain discussions. This book sets the future direction for employability discussions in South Africa and beyond with the advent of the 4IR. According to Linstone and Turoff (2002), the Delphi technique can also be adopted in one or more of the following scenarios:

1   If the problem of the subject matter does not align with precise analytical techniques but can be collectively deliberated upon based on subjective opinions.
2   When the research study contributes to the investigation of a specific problem with no prior history of sufficient exploration and may get opinions of experts from diverse backgrounds and perspectives, which this current research considers.
3   When there is limited time and cost to organize group meetings regularly to conduct the research study.
4   When organizing a structured communication process to increase the efficiency of face-to-face (physical) meetings.
5   The need for experts to remain anonymous to ensure the validity of results to avoid 'bandwagon effect'.

(Linstone and Turoff, 2002)

Components of the Delphi method

A typical Delphi technique consists of five major components as described by Loo (2002). These five characteristics are adopted in this present study:

1   A well-selected panel of experts who are well informed on the specific topic or subject in question should be considered for the study.
2   The panel of experts should remain anonymous throughout the study.

3　During the Delphi process, it is expected that the researcher designs a structured questionnaire and feedback report for the panel of experts.

4　An iterative process that involves multiple rounds to determine when consensus is reached by the panel of experts.

5　The designing of a research report which reflects the results of the Delphi study which provides forecasts, action plans, and recommendations.

Furthermore, six guidelines for using Delphi technique were suggested by Hasson *et al.* (2000). According to them, the following must be considered during the Delphi design:

1　*Identifying the research problem:* According to Turoff (1970), the Delphi technique is considered when an informed perspective on a subject matter is required from a broad range of disciplines, hence the use of experts. According to McKenna (1994), engaging experts to share their knowledge in a subject matter is appropriate, especially if previous studies have not adopted a similar technique. This current employability study goes one step better in utilizing the Delphi technique as a tool of investigation. This technique also examines the validity of the cross-disciplinary nature of the subject matter or issue in question.

2　*Proper understanding of the Delphi process:* One of the merits of the Delphi technique is generating consensus after a multistage process of collecting and analysing opinions from the experts (McKenna, 1994).

3　*Experts' selection:* This is the process of shortlisting and selecting the panel members who are interested in the topic, knowledgeable, committed to responding to multiple rounds of questions, and impartial. Depending on the scope of the study, some researchers can accept more than 100 experts or as few as 10. An elaborate explanation on how the experts were chosen for this study will be presented in this chapter of the book.

4　*Communication with experts:* This deals with the process of explaining key information to the experts. This includes the amount of time required, instructions before filling, the objectives of the study, and how the information will be utilized for the research study.

5　*Data collection and analysis:* This deals with collecting and recording expert opinions. The Delphi technique can go through several rounds, although two or three rounds are preferable with at least an 80% consensus rate (stability or acceptance of data) as the minimum target (Green *et al.*, 1999). Analytical software was adopted to analyse the various responses (rounds) to provide experts with feedbacks on central tendencies – median (M) and interquartile range (IQD) – as well as levels of dispersion – standard deviation.

6　*Presenting and interpreting results:* This deals with the presentation of data from the Delphi study. Notably, two methods are prominent: graphical and statistical. Both methods were adopted in the realization of this book.

Given the nature of this study, the Delphi technique is appropriate to obtain credible inputs from experts in the construction industry and academia to serve as a key input in the development of an employability improvement framework for built environment graduates.

Executing the Delphi study

In the successful execution of the Delphi technique, the following stages are pertinent according to Skulmoski *et al.* (2007):

1   *The formulation of research questions and objectives:* These can be constructed on the basis of the researcher's industry experience or a comprehensive literature review to determine whether theoretical gaps actually exist.
2   *Research design:* This involves revising the various research methods and methodologies and deciding which of them is most suitable to answer the research questions. The Delphi method comes in handy when expert opinions in a particular subject matter are required.
3   *Research sampling:* The selection and shortlisting of experts is a critical component of the Delphi method because their knowledgeable opinions determine the output of the Delphi process. In selecting experts, there is a pertinent need to consider their knowledge and experience in the subject matter under focus. Their capacity, available time, and willingness to participate in the study are also key factors to consider when selecting experts.
4   *Formulating the first round of the Delphi questionnaire:* The formulation of the initial Delphi questions are significant and these must be constructed clearly and coherently. This is because the Delphi output is largely dependent on the manner of the questions and the way they are framed by the researcher. Hence, to obtain credible and appropriate responses, the Round One questionnaire must be developed cautiously.
5   *Formulating a pilot study:* A pilot study is required to evaluate the feasibility and coherence of the Delphi questionnaire. In most cases, the pilot study helps to improve the overall credibility of the questions, to check grammar, and to resolve any structural problems with the questions.
6   *Analysing the first round of the Delphi questionnaire:* This involves distributing the questionnaires to selected experts who complete and return them to the researcher who analyses them based on central tendencies – median (M) and interquartile range (IQD) as well as levels of dispersion – standard deviation. Continuous verification by the researcher is required to improve the reliability of the study.
7   *Formulating the second round of the Delphi questionnaire:* The responses obtained from the first round of the Delphi questionnaire form the basis for the second round. The second round focuses mostly on the opinions of experts during the first round.
8   *Analysing the second round of the Delphi questionnaire:* This involves distributing the Round Two questionnaires to selected experts who complete and return them to the researcher who analyses them. In this round, the experts are given the opportunity to verify their initial responses, to retain their

initial opinions (choices), or to change their initial opinions because the results of other experts are shared in the latest round. Again, constant and careful verification by the researcher is required to improve the consistency and reliability of the results

(Skulmoski *et al.*, 2007).

In cases where consensus is not reached after the first two rounds, a third round of the Delphi questionnaire is required. Figure 8.1 shows the Delphi process for this study. Moreover, for the successful planning and execution of the Delphi study, the following four key sequential processes are considered:

1   Problem definition and question formulation
2   Panel selection
3   Panel size determination
4   Delphi iterations

(Loo, 2002)

Problem definition and question formulation

The construction of Delphi questions is important to the entire process. The panel of experts must understand the whole context of the questionnaire to provide the appropriate answers. To achieve the objectives of the Delphi study, key questions were designed to get relevant information from experts. The basis for the formulation of these questions is presented in Table 8.1.

Selection of Delphi experts

The successful execution of the Delphi interview process is largely dependent on the quality and the knowledgeability of experts providing responses and opinions (Hasson *et al.*, 2000; Hsu and Sandford, 2007). For this book, an expert or a group

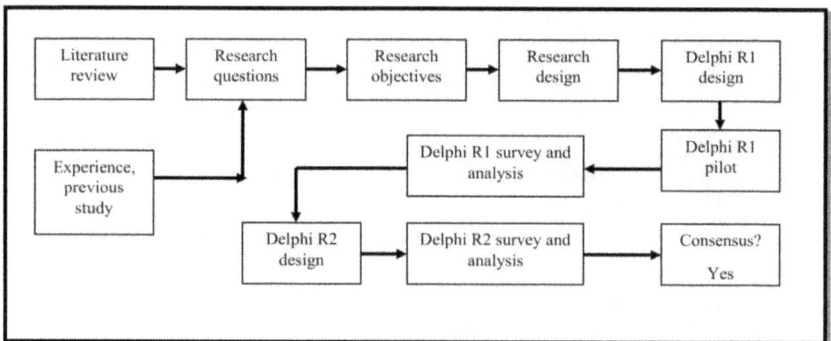

Figure 8.1 Delphi design for this study

Table 8.1 Basis for formulating the Delphi questions

Key Delphi questions	Study consideration
Why the Delphi study?	Owing to the dynamism of the construction industry, employers' expectations of graduate skills are ever-changing. With the advent of the fourth industrial revolution, employers' expectations have gone even higher, placing significant pressures on universities
What is needed to be understood that previously was not?	Factors and variables that contribute to the employability of built environment graduates in the fourth industrial revolution
How will Delphi results impact this study?	The results of the Delphi study will be used to develop an employability improvement framework for built environment graduates

of experts refer to a panel of knowledgeable individuals who were selected based on criterion sampling (Okoli and Pawlowski, 2004). It is also germane that the experts were willing to participate in the process to ensure a high commitment response rate. Over the years, researchers have exercised flexibility in establishing criteria in deciding who should be called an expert. According to Veltri (1986), an expert should possess at least one of the following: a robust understanding of the subject matter under focus; authorship of peer-reviewed publications related to the subject or discipline in question; and regular participation in conferences and workshops related to the subject or discipline in question. The flexibility of these guidelines may lead to a large number of individuals who qualify as experts, hence the quality of the research results may be compromised. In the same vein, Adler and Ziglio (1996) recommend four key requirements for Delphi experts. According to them, they must be knowledgeable and experienced in the subject matter under focus; willing to participate in the study; possess ample time to participate in the study and partake in several rounds if required; and be good communicators to express their suggestions and opinions with minimal issues and clarity. Rogers and Lopez (2002) suggested that in a typical Delphi study, experts must satisfy at least two of the following requirements. These include authoring a peer-reviewed publication; presentation at a conference; chaired or is a member of a committee; practising (employed) with at least five years of relevant experience; and employed as a staff member with an accredited institution of higher education. In addition, Hallowell and Gambatese (2010) recommended that experts must satisfy at least four of the following: authorship of several peer-reviewed articles relating to the subject matter; presentation at a conference; chair or member of a committee; at least five years of relevant experience in the construction industry; faculty member at an accredited institution of higher education; advanced academic qualification in a related discipline; and registered with a professional body. As observed, researchers have the autonomy to design criteria as they deem fit to suit the purpose of their study.

In selecting the panel of experts for this current research study, the recommendations of both Rogers and Lopez (2002) and Hallowell and Gambatese (2010) were considered. As shown in Table 8.2, an expert for this study's Delphi process was required to satisfy three or more of the five listed criteria (Chan *et al.*, 2001):

1   Possess at least a bachelor's degree in a discipline within the built environment.
2   Currently employed with a registered and recognized tertiary institution in South Africa or professional in the construction industry.
3   At least five years as a staff member with any registered and recognized tertiary institution in South Africa or at least five years of working experience in the construction industry.
4   Affiliated with recognized professional bodies so that their opinions, suggestions, and recommendations may be applicable and transferable to the wider population.

Table 8.2 Assessment of Delphi expert qualifications

S/N	Eligibility criteria for experts	E1	E2	E3	E4	E5	E6	E7	E8	E9	E10	E11	E12	E13	E14
1	Possess at least a bachelor's degree	X	X	X	X	X	X	X	X	X	X	X	X	X	X
2	Currently employed with a tertiary institution or professional in the construction industry	X	X		X			X	X	X	X	X		X	X
3	At least five years of working experience with a tertiary institution or construction industry	X	X	X	X	X	X	X	X	X	X	X	X	X	X
4	Affiliated with professional bodies	X	X	X	X	X	X	X		X	X	X	X	X	
5	Author or co-author of a peer-reviewed publication	X	X	X	X		X	X	X		X		X		X
	Total	5	5	4	5	3	4	5	4	4	5	4	4	4	4

Source: Author's compilation.

5   Author or co-author of a peer-reviewed publication in any of the disciplines within the built environment and have presented at conferences and workshops.

As seen in Table 8.2, the experts selected for this study all satisfied three or more of the listed criteria. Experts were recruited via an email, which described the objectives of this research study with 22 invitations sent out initially. Subsequently, 17 responded and agreed to the initial invitation. These 17 respondents were sent a comprehensive description of what the Delphi study entails. The potential experts were asked to send their curriculum vitae (CV) to ascertain their study field and area of expertise and to verify whether they met the study's qualification standards. After this verification and confirmation process, the selected experts received the first round questionnaire survey. The first round survey consisted of both open-ended and closed questions with the experts required to mark and state their opinions. Seventeen experts successfully completed the first round of the questionnaire survey. However, 3 members dropped out despite several reminders, allowing 14 members to successfully complete the survey rounds. The three members who failed to respond could have withdrawn owing to the rigorous nature and time-consuming tendencies of the Delphi method. Based on previous recommendations from different scholars, the number of active experts (14) who completed the survey rounds was deemed adequate.

This was echoed by Delbecq *et al.* (1975), who confirm that 10–15 experts are adequate if the backgrounds of the panel members are similar, something which this study achieved. In addition, Rowe *et al.* (1991) suggest that in peer-reviewed studies, the Delphi panel size can range from 3 to 80 members, hence the number adopted by this study is suitable. According to Skulmoski *et al.* (2007), the Delphi panel size can range between 10 and 18 members, while Hallowell and Gambatese (2010) insist on a minimum of eight (8) members as appropriate. Hence, the number of experts (14) adopted for the Delphi process in this current research study is suitable because the Delphi process is based on arriving at a consensus and not statistical influence or power. Furthermore, the Delphi panel size should be based on several factors such as the choice of geographical location, the capacity of the researcher, characteristics of the study in focus, and the number of willing and available experts (Hallowell and Gambatese, 2010). A drop of only 3 members from the first round with 14 completing the Delphi process further shows how the issue of graduate employability in the era of the 4IR is both engaging and critical in South Africa.

Panel size determination

Because the Delphi technique tends towards a qualitative approach rather than a quantitative one, fewer experts are required to participate in the study when compared to normal quantitative surveys. However, establishing the minimum number of experts for Delphi studies has been debated over the years and various researchers have suggested various sample sizes. A panel of seven experts was adopted by Dalkey and Helmer (1963) and supported by Linstone and Turoff (2002). Cavalli-Sforza and Ortolano (1984) assert that a typical Delphi panel should have 8–12 experts. This is similar to Phillips (2000) who recommends between 7 and 12 experts. In cases where the group of experts is homogeneous (belonging to similar backgrounds and share similar views and characteristics), 10–15 participants are recommended (Andranovich, 1995; Skulmoski et al., 2007). On the other hand, if the group of experts are of different opinions and characteristics (an international study), then a larger sample is required to maintain balance (Skulmoski et al., 2007; Zami and Lee, 2009). It is also necessary to state that larger samples present issues with data collection and analysis as well as challenges in reaching consensus (Skulmoski et al., 2007). Another significant factor is the availability of experts to participate. In cases where experts are limited, a smaller sample is expected (Lam et al., 2000). For this current research study, owing to time constraints and varying schedules of experts, a relatively small sample was adopted and subsequently a follow-up research was conducted through a questionnaire survey. Hence, a sample size of 17 experts was adopted based on certain criteria that were discussed earlier. Owing to the nature of the subject area, experts were expected to represent a broad spectrum of backgrounds (fairly split between academics and industry) to provide knowledgeable perspectives to the study.

Delphi iterations

Another significant aspect of the Delphi process is the number of iterations (series of rounds) before reaching consensus (Hsu and Sandford, 2007). Woudenberg (1991) asserts that two to ten rounds are appropriate and believes that accuracy increases over the course of more iterations owing to the replication of opinions and general pressure for conformity. In the same vein, Critcher and Gladstone (1998) suggest between two and five rounds. For most studies, two or three iterations are satisfactory (Skulmoski et al., 2007). In cases where the sample is heterogeneous, three or more iterations are recommended. For this current research, the expert panel was homogeneous because they all possessed a background in the built environment (both academia and industry). Hence, the consensus was achieved after two rounds of iteration with each round taking an average of a month to complete.

The first round questionnaire was designed based on the literature review to determine the various dimensions (main and sub-dimensions) that contribute to the employability of built environment graduates in South Africa. The first round questionnaire was emailed to 17 experts with 14 experts responding within the stipulated time. However, three experts failed to respond despite reminder emails and an additional week. The experts' responses and opinions were presented and analysed on Microsoft Excel spreadsheet and the median, mean, standard deviation, and the interquartile values were observed. In the second round, a questionnaire was formulated based on the responses and opinions of the first round. The second round presented an opportunity for experts to review, rank, and make comments on the various dimensions (main and sub-dimensions) that contribute to the employability of built environment graduates which were addressed by the experts in the first round. Moreover, the questions in the second round were close-ended which provided the opportunity for experts to agree, disagree, or seek clarifications concerning the outlined dimensions that contribute to improved graduate employability. Based on recommendations by some experts, identical factors were reworded (Chan *et al.*, 2001). This was achieved using a ten-point Likert scale of 'no significance', 'low significance', 'medium significance', 'high significance', and 'very high significance', while 'no significance' was assigned the lowest weighting (1 and 2), 'very high significance' had the highest weighting (9 and 10). The degree of consensus reached amongst experts regarding the outlined dimensions that contribute to improved graduate employability were measured by the frequencies obtained. Based on analyses of the second round, the dimensions that contribute to improved graduate employability were prepared, which conceptualizes the framework for the broader study.

For this research study, the mode of communication was well thought out and considered the possible issues surrounding the anonymity of experts and the choice of analysis. There are two key forms of interaction: a paper-based mode and an electronic medium (Hasson *et al.*, 2000). Unlike the paper-based mode, the electronic medium (emails) offers timely and swift communication channels (real-time messaging). This immediate response system provided by emails makes this mode of communication more convenient than the paper-based mode as it sustains the interest of the experts. Hence, the Delphi survey for this research was successfully carried out via email and follow-up thread messages were used to remind experts and provide immediate clarification where necessary.

Evaluation criteria

At the end of each round, the study established some criteria to analyse responses from the experts. For this study, each round required a minimum of three weeks to afford experts ample time to respond to the questions. Because the first round was mainly an exploratory round, experts (based on their knowledge and experience) were instructed to indicate the significance of certain factors across various degrees of measurement (extent) and to provide extra factors where necessary. For the second round, experts were instructed to either accept the group median, maintain original opinions, or record a new response. After this second round, consensus was reached. Before the Delphi survey commenced, the questions were approved by the researcher's supervisor and piloted with different experts to ascertain whether the objectives of the research could be realized with the structured questionnaire. The pilot study also improved the overall comprehension of the questions. The pilot team constituted five experts (three from academia and two from the industry) and fellow PhD students.

Computing data from the Delphi study

As stated earlier, experts' responses and opinions were presented and analysed on Microsoft Excel spreadsheets with columns assigned for the median, mean, standard deviation, and interquartile values. Each question element and factor contributing to the employability of built environment graduates in South Africa and other related issues was presented. For every round, the respective group median value and interquartile deviation (IQD) were computed as a measure of the central tendencies to determine consensus. Owing to its tendency to minimize the influence of potentially biased persons, the median value was adopted as a measure of central tendency as opposed to the mean and IQD. On the other hand, the respective IQD scores helped to summarize data variability and hence identified the appropriate measures that influenced graduate employability. By eliminating outlying values, the IQD also offers general coherence and clarity of the overall dataset. The interquartile range, which is closely related to the median, indicates the extent to which the central 50% of values within the presented dataset are dispersed. After the second round of the Delphi, Equation 8.1 was adopted to calculate the absolute deviations (Di) of the group medians ($m(X)$) of each rating or ranking for the respective questions:

$$Di = \left[xi - m(X) \right] \hspace{4cm} \text{Equation 8.1}$$

Here,
Di represents absolute deviation
xi represents panel ratings
$m(X)$ represents the measure of central tendency

Determining consensus from the Delphi study

After the first two rounds, the group median and the IQD were computed for all responses and consensus was reached on all questions. In addition, the respective deviations for all responses were calculated using the absolute median, as shown in Equation 8.1. In determining consensus, several researchers have adopted numerous parameters and judgements as benchmarks. Holey *et al.* (2007) state that consensus refers to an agreement of opinions and can be determined by the following parameters:

1   The summative judgements and opinions made by experts.
2   Opinions from experts shifting towards central tendency.
3   By checking the consistency and steadiness of experts' responses between subsequent rounds of the Delphi study.

According to Rayens and Hahn (2000), consensus can be reached by considering both the mean values and standard deviations values where a decrease in the latter between subsequent rounds shows high agreement levels. Rayens and Hahn (2000) further suggest the adoption of the IQD in determining the consensus, a parameter adopted by this study. Rayens and Hahn (2000) further recommended another criterion to achieve the stability of results. They state that the IQD should equal 1, meaning over 60% of the experts provided responses that were generally positive or generally negative with regard to the subject matter. In the same vein, factors with an IQD value of less than 1 suggest that responses that were generally positive or generally negative fell between 40% and 60% which signified consensus. This study by Rayens and Hahn (2000) resonates with that of Raskin (1994), who maintains that an IQD value of one 1 or less can be construed as an indicator of consensus. Furthermore, Holey *et al.* (2007) provide several parameters to confirm whether consensus has been reached:

1 A significant increase in percentage agreements.
2 When importance rankings achieve convergence.
3 A significant increase in Kappa values.
4 Reduced or no comments and responses from experts as rounds progress.
5 Significant reduction of standard deviation values.

Based on the foregoing parameters, consensus can be reached when there is a convergence of ideas and opinions from the experts. For this current study, consensus was achieved when the following were observed:

1 When over 60% of responses for each question were ranked generally positive or generally negative.
2 When the average of the absolute deviation as calculated from Equation 8.1 was not more than 1 unit.
3 When the IQD had a value that was less than 1, meaning factors with IQD = 0.00 were considered to have a high agreement and consensus among experts.

Consequently, the consensus scales for this study are shown in Table 8.3.

Validity and reliability of the Delphi study

Both validity and reliability are necessary for studies adopting a qualitative approach as they check the consistency of the study's research design and paradigms across various studies (Creswell, 2014). Reliability deals with the extent to which a well-thought-out process provides comparable and identical results

Table 8.3 Consensus scales for this study

S/N	Consensus strength	Median	Mean	Interquartile deviation (IQD)
1	Strong	9–10	8–10	≤ 1 and $\geq 80\%$ (8–10)
2	Good	7–8.99	6–7.99	$\geq 1.1 \leq 2$ and $\geq 60\% \leq 79\%$ (6–7.99)
3	Weak	≤ 6.99	≤ 5.99	$\geq 2.1 \leq 3$ and $\leq 59\%$ (5.99)

under constant conditions at every point in time. However, this is difficult to accomplish in Delphi-based studies as decisions and findings are based on diverse perspectives of experts based on their knowledge, experience, and perceptions towards ideas. In qualitative studies, validity (internal and external) remains an important consideration, and this current research made a conscientious effort to ensure the accuracy and integrity of the study findings from the researcher's perspective. As a measure towards ensuring credibility, the study also considered the response consistency from experts.

The validity of the research was further boosted by ensuring that experts were completely anonymous from each other to eliminate the 'bandwagon' effect. By establishing several rounds in the Delphi study, experts were given a chance to maintain their opinions or make changes whilst providing explanations for their arguments and differing views. Hence, the iteration process enhanced the internal validity of the study. In addition, the back-and-forth email communication between the researcher and the panel of experts further ensured the internal validity of this study. Meanwhile, the reliability of the Delphi study was enhanced through the selection of experts from a homogeneous background (possessing built environment orientation). Experts were considered from both academia and industry to improve the balance of opinions (Creswell, 2014). To further improve the reliability and credibility of the research study, there was strict adherence to the qualification criteria across the board. Based on the foregoing discussions, this study fulfilled requirements for both validities (internal and external) and reliability in conformity with standard research ethics.

Delphi study findings

Delphi experts' information

For this study's Delphi, 14 members successfully completed the survey rounds. As stated earlier, experts' responses and opinions were presented and analysed on Microsoft Excel spreadsheet with columns assigned for the median, mean, standard deviation, and the interquartile values. For this study, experts were required to possess at least a bachelor's degree in a discipline within the built environment. The qualifications held by the experts are shown in Table 8.4. Five of the experts had a doctor of philosophy (PhD) degree, seven of the experts had a

Table 8.4 Panel of experts' qualifications

Highest qualification	Number of experts
Doctor of Philosophy (PhD)	5
Master's degree (MSc and MEng)	7
Bachelor's degree (BEng)	2
Total	**14**

master's degree (master of science [MSc] and master of engineering [MEng]), and two had a bachelor's degree.

As seen from Table 8.4, 86% of the experts possessed postgraduate degrees (master's and PhD) which explains the high quality of their responses. In the discussions surrounding graduate employability, the opinions of experts (from both academia and industry) with high academic qualifications further improve the credibility of this study. Experts with a master's or PhD degree in the built environment have a robust understanding of what is required of students to thrive in the world of work after graduation. Hence, the high qualifications of the experts helped to ensure the high quality and reputability of this study's Delphi process. Experts were also required to be currently employed with a registered and recognized tertiary institution in South Africa or be a professional in the construction industry. In terms of their field of specialization, all experts were currently active in the built environment in various capacities in both academia and industry, as shown in Table 8.5. From their CV analysis, the experts adequately represented most of the various disciplines in the built environment, thereby improving the reliability of the study. Moreover, eight of the experts were from higher institutions, while six were actively practising in the construction industry.

Experts were further required to possess at least five years' experience as a staff with any registered and recognized tertiary institution in South Africa or at least five years of working experience in the construction industry. As seen from Table 8.6, one expert had 1–5 years of experience, seven experts had 6–10 years of experience, three experts had 11–20 years of experience, two experts had 21–30 years of experience, and one expert had more than 31 years of experience.

Table 8.5 Panel of experts' field of specialization

Field of specialization	Number of experts
Architecture	2
Quantity surveying	1
Construction project management	2
Engineering (civil, mechanical, electrical)	9
Total	**14**

Table 8.6 Panel of experts' years of experience

Years of experience	Number of experts
1–5	1
6–10	7
11–20	3
21–30	2
Over 31 years	1
Total	**14**

Experts were further required to be registered and affiliated with recognized professional bodies so that their opinions, suggestions, and recommendations may be applicable and transferable to the wider population. Five of the experts were registered with the Engineering Council of South Africa (ECSA), two were registered with the South African Council for the Project and Construction Management Professions (SACPCMP), two were registered with the South African Institution of Civil Engineering (SAICE), and one was registered with Project Management South Africa (PMSA). Finally, experts for this study were further required to be authors or co-authors of peer-reviewed publications in any of the disciplines within the built environment as well as have presented at conferences and workshops. Most of the experts had published books and monographs, book chapters, peer-reviewed academic journal articles, or conference papers. Remarkably, one of the experts had a patent project for effective land use. Six of the experts had previously served on editorial boards of several peer-reviewed journals and conference proceedings.

Main constructs of the employability framework

From the review of existing employability models, frameworks, and their corresponding constructs across several contexts, this book identified six key dimensions (constructs or critical factors) in ensuring the improvement of graduate employability. It is upon these dimensions that the conceptual employability improvement framework for this study was achieved. As discussed earlier in this book, one of the critical dimensions that contribute to the employability of students is the possession of generic skills. Considering the various dynamics and complexities of the industry, including increased global competition, technological advancements, environmental factors, and construction projects sophistication, the need for graduates to possess generic skills cannot be overstated. These changes require graduates to possess not only a sound academic background but also non-academic skills and competencies.

Another key dimension identified from the various models are discipline-specific skills that students have to possess during their academic cycle. Generally, the academic degree is the primary consideration by most employers when making their recruitment decisions. This is because they take into consideration how and when students successfully complete their academic courses or disciplines. The importance of work experience in improving the employability of built environment graduates cannot be quantified. Hence, its inclusion in this study. By engaging in work activities, built environment students are presented with enhanced knowledge about their career choices and related occupations. Hence, these opportunities help students ascertain the strengths, weaknesses, and expectations of their chosen fields and how they can explore other possible options. For this study, the term work-integrated learning (WIL) was adopted.

Another key dimension is emotional intelligence which is the capacity to recognize one's feelings and that of others. By possessing self-confidence, self-management, self-awareness, social awareness, and internal locus of control, graduates can successfully adapt to varying work conditions. Also, in enhancing the employability of graduates for the industry, the collaboration between universities and industry plays a key role and its importance cannot be ignored. With the present-day industry driven by various factors including technology and globalization, it is imperative for universities to establish cordial connections with industry counterparts in a bid to provide opportunities for students to learn more about these significant changes. In embracing the 4IR, built environment graduates will require a plethora of skills as automation continues to advance. Some of these skills and attributes include the ability to determine the deeper meaning of innovations (sense-making skills), the ability to connect with others deeply and directly (social intelligence), the ability to operate and adapt to different cultural settings (cross-cultural competence), and proficiency in proffering solutions (adaptive thinking), amongst others. Hence, the need for built environment students to possess 4IR skills and knowledge.

As the fundamental constructs for the proposed employability improvement framework for built environment graduates, these six dimensions were included in the first round of Delphi to obtain the opinions of experts. This is shown in Table 8.7. Owing to its tendency to minimize the influence of potentially biased persons, the median value was adopted as a measure of central tendency as opposed to the mean and IQD. This was achieved using a ten-point Likert scale of 'no importance', 'low importance', 'medium importance', 'high importance', and 'very high importance'.

As shown in Table 8.7, Delphi experts were asked to state the level of importance of the following employability dimensions on graduate employability. Generic skills earned an absolute score of 10, indicating its very high importance on graduate employability. Hence, the need for graduates to possess generic skills cannot be overstated. Discipline-specific skills that are obtained by virtue of studying a course or discipline received a score of 9, highlighting its high importance. The same score (9) was recorded for work-integrated learning, emotional intelligence, and university–industry collaboration (UIC) alike. Remarkably, Industry 4.0 knowledge also attracted an absolute score of 10, confirming its very high importance regarding graduate employability in the era of the 4IR. Hence, the experts for this study were in full agreement with the adoption of these six dimensions in developing an employability improvement framework for built environment graduates.

Table 8.7 Experts' responses to main dimensions

Employability dimensions	Median
Generic skills	10
Discipline-specific skills	9
Work-integrated learning	9
Emotional intelligence	9
University–Industry collaboration (UIC) outcomes	9
Industry 4.0 requirements	10

Generic skills (first and second round Delphi results)

During the first round of the Delphi process, a total of 16 generic skills were identified from the literature review, as shown in Table 8.8. Experts were required to indicate the significance of the generic skills in contributing to graduate employability. The rating was achieved using a 10-point Likert scale of 'no significance', 'low significance', 'medium significance', 'high significance', and 'very high significance'. While 'no significance' was assigned the lowest weighting (1 and 2), 'very high significance' had the highest weighting (9 and 10). At this stage, IQD above 1.00 were recorded which necessitated a second round.

However, one of the experts suggested two additional skills: interdisciplinary skills and transformative skills. Both were included in the second round and experts were requested to rank their level of significance. After the second round, a consensus was achieved, as shown in Table 8.9.

As discussed earlier, a strong consensus is reached when factors have a median value between 9 and 10, a mean value between 8 and 10, and an IQD value = 1 or less than 1. As seen from Table 8.9, *critical-thinking skills, creativity/innovation skills, willingness to learn, information and communication technology (ICT) skills*, and *problem-solving skills* achieved very good consensus. This implies that these five sets of generic skills have the greatest significance on graduate employability. Apart from these 5 generic skills, the other 13 skills were ranked highly with median values of 8. This implies that these 13 skills have an equal significance on graduate employability. The consensus was further measured by inspecting IQD

Table 8.8 Generic skills factors (first round results)

Generic skills	Median	Mean ($\bar{x}$)	SD (σX)	IQD
Communication skills	8	8.35	1.69	2.00
Critical thinking skills	9	8.47	1.59	2.00
Teamwork skills	8	8.24	1.44	1.00
Interpersonal skills	8	8.12	1.62	2.00
Leadership skills	8	8.12	1.50	2.00
Adaptability skills	8	7.94	1.43	2.00
Creativity/innovation skills	9	8.29	1.31	2.00
Willingness to learn	9	8.82	1.29	2.00
Coordinating/organizing skills	8	7.88	1.93	2.00
Management skills	8	7.88	2.03	2.00
ICT skills	9	8.65	1.50	2.00
Decision-making skills	8	8.12	1.41	2.00
Analytical skills	8	8.35	1.37	3.00
Problem-solving skills	9	8.71	1.26	3.00
Entrepreneurship skills	8	7.65	2.18	3.00
Networking skills	8	7.76	2.17	4.00

Table 8.9 Generic skills factors (second round results)

Generic skills	Median	Mean ($\bar{x}$)	SD (σX)	IQD
Communication skills	8	7.79	1.25	0.50
Critical thinking skills	9	8.57	1.34	0.00
Teamwork skills	8	7.79	1.05	0.00
Interpersonal skills	8	7.71	1.27	0.00
Leadership skills	8	7.86	1.03	0.00
Adaptability skills	8	7.86	0.86	0.50
Creativity/innovation skills	9	8.79	0.89	0.00
Willingness to learn	9	8.86	0.66	0.50
Coordinating/organizing skills	8	7.86	1.29	0.00
Management skills	8	7.64	1.45	0.50
ICT skills	9	8.71	1.27	0.00
Decision-making skills	8	8.00	0.88	0.00
Analytical skills	8	8.07	0.73	0.00
Problem-solving skills	9	8.86	0.86	0.00
Entrepreneurship skills	8	7.71	1.59	0.50
Networking skills	8	7.79	1.48	0.00
Interdisciplinary skills	8	8.07	0.83	0.00
Transformative skills	8	8.00	1.30	0.00

values, as seen in Table 8.9. As previously stated, factors with IQD value less than 1 were considered to have high agreement and consensus among experts. All 18 satisfied these criteria; hence consensus was fully achieved in the second round.

Discipline-specific skills (first and second round Delphi results)

During the first round of the Delphi process, a total of 18 discipline-specific skills were identified from the literature review, as shown in Table 8.10. Experts were required to indicate the significance of the discipline-specific skills in contributing to graduate employability. The rating was achieved using a 10-point Likert scale of 'no significance', 'low significance', 'medium significance', 'high significance', and 'very high significance'. While 'no significance' was assigned the lowest weighting (1 and 2), 'very high significance' had the highest weighting (9 and 10). At this stage, IQD above 1.00 were recorded which necessitated a second round.

However, one of the experts suggested three additional skills: collaborative skills, transformative skills, and hands-on experience. All three were included in the second round and experts were requested to rank their level of significance. After the second round, consensus was achieved, as shown in Table 8.11.

As seen from Table 8.11, both *knowledge of the subject area* and *lifelong learning* possess a strong median value of 10 and mean values of 9.33 and 9.44, respectively, indicating very good consensus. This implies that both factors have the greatest

Table 8.10 Discipline-specific skills factors (first round results)

Discipline-specific skills	Median	Mean ($\bar{x}$)	SD (σX)	IQD
Knowledge of the subject area	10	9.00	1.20	2.00
Knowledge of curricula	9	8.27	1.75	2.00
Realizes strength and weaknesses	8	7.93	1.59	2.00
Provides critical thinking	9	8.40	1.25	2.00
Pedagogical skills	8	8.00	2.21	1.00
Organizational skills	8	7.80	1.85	2.00
Communication skills	9	8.53	1.74	2.00
Administrative skills	9	8.00	2.03	2.00
Management skills	9	8.13	1.52	2.00
Improved motivation	8	8.27	1.50	1.00
Critical thinking skills	9	8.87	1.25	2.00
Lifelong learning	10	9.14	1.66	2.00
Decision-making skills	9	8.79	0.89	1.00
Digital literacy	9	8.47	1.42	2.00
Develop relationships with peers	8	7.60	1.55	3.00
Self-confidence skills	8	8.20	1.30	1.00
Connection with other disciplines	8	8.20	1.50	2.00
Problem-solving skills	9	8.87	0.99	1.00

Table 8.11 Discipline-specific skills factors (second round results)

Discipline-specific skills	Median	Mean ($\bar{x}$)	SD (σX)	IQD
Knowledge of the subject area	10	9.33	1.09	0.00
Knowledge of curricula	9	8.78	1.01	0.00
Realizes strength and weaknesses	8	7.78	0.95	0.00
Provides critical thinking	9	8.78	0.73	0.50
Pedagogical skills	8	8.11	1.68	0.00
Organizational skills	8	7.56	1.42	0.00
Communication skills	9	8.33	1.61	0.00
Administrative skills	9	8.44	1.50	0.00
Management skills	9	8.67	1.09	0.00
Improved motivation	8	7.56	1.22	0.00
Critical thinking skills	9	8.56	1.07	0.00
Lifelong learning	10	9.44	1.76	0.00
Decision-making skills	9	8.78	0.70	0.00
Digital literacy	9	8.67	1.34	0.00
Develop relationships with peers	8	7.67	1.37	0.00
Self-confidence skills	8	7.78	0.95	0.00
Connection with other disciplines	8	7.78	1.37	0.00
Problem-solving skills	9	8.89	0.66	0.00
Collaborative skills	8	8.33	0.83	0.50
Transformative skills	8	7.89	1.07	0.50
Hands-on experience	9	9.11	0.55	0.50

significance on graduate employability. Apart from this, a total of 10 other factors possessed a strong median value of 9, which indicates an equal significance on graduate employability. Consensus was further measured by inspecting IQD values, as seen in Table 8.11. As previously stated, factors with IQD value less than 1 were considered to have high agreement and consensus among experts. All 21 discipline-specific skills satisfied these criteria, hence consensus was fully achieved.

Work-integrated learning factors (first and second round Delphi results)

During the first round of the Delphi process, a total of 28 work-integrated learning factors skills were identified from the literature review, as shown in Table 8.12.

Table 8.12 Work-integrated learning factors (first round results)

Work-integrated learning factors	Median	Mean ($\bar{x}$)	SD (σX)	IQD
Developing practical knowledge	9	8.59	1.80	2.00
Understand job responsibilities	9	8.29	1.31	1.00
Awareness of industry	9	8.18	1.51	1.00
Improved work attitude	8	8.00	1.32	1.00
Develops interpersonal values	8	7.71	1.65	2.00
Résumé development	8	7.76	1.99	2.00
Awareness of workplace culture	8	7.94	1.56	3.00
Understanding professional ethics	8	8.47	1.18	1.00
Time management	9	8.47	1.37	2.00
Digital literacy	8	7.88	1.80	2.00
Self-confidence	8	7.88	1.76	2.00
Developed professional identity	8	8.12	1.54	1.00
Problem-solving ability	8	8.29	1.16	1.00
Knowledge of contemporary issues	8	7.76	1.52	2.00
Respect for oneself	8	7.82	1.85	2.00
Knowledge of safety-procedures	8	7.94	1.68	2.00
Working in multidisciplinary teams	8	8.24	1.25	2.00
Knowledge of engineering designs	8	8.18	1.01	1.00
Knowledge of quality control measures	8	7.59	1.91	2.00
Site-supervision knowledge	8	7.47	1.84	2.00
Financial management skill	8	7.59	1.91	3.00
Interpersonal relationship with industry professionals	8	7.53	1.66	2.00
Networking with other interns on-site	8	7.41	1.62	3.00
Exposure to career opportunities	9	8.06	1.64	2.00
Enhanced learning	9	8.29	1.69	3.00
Acquisition of professional skills	9	8.06	1.98	3.00
Technical report writing	8	7.65	1.54	3.00
Mentorship opportunities	8	7.59	1.80	3.00

Experts were required to indicate the significance of each factor in contributing to graduate employability. The rating was achieved using a 10-point Likert scale of 'no significance', 'low significance', 'medium significance', 'high significance', and 'very high significance'. While 'no significance' was assigned the lowest weighting (1 and 2), 'very high significance' had the highest weighting (9 and 10). At this stage, IQD above 1.00 were recorded which necessitated a second round.

No new additions were made during the first round and consensus was achieved after the second round, as shown in Table 8.13.

As seen from Table 8.13, *time management, developing practical knowledge, understand job responsibilities, awareness of industry, exposure to career opportunities, enhanced learning,* and *acquisition of professional skills* all possess a strong median

Table 8.13 Work-integrated learning factors (second round results)

Work-integrated learning factors	Median	Mean ($\bar{x}$)	SD (σX)	IQD
Developing practical knowledge	9	8.57	1.60	0.00
Understand job responsibilities	9	8.64	1.22	0.00
Awareness of industry	9	8.57	1.45	0.00
Improved work attitude	8	7.79	1.19	0.00
Develops interpersonal values	8	7.71	1.44	0.00
Résumé development	8	7.71	1.44	0.00
Awareness of workplace culture	8	7.86	0.86	0.00
Understanding professional ethics	8	8.14	0.66	0.50
Time management	9	8.71	0.99	0.00
Digital literacy	8	7.79	1.37	0.00
Self-confidence	8	7.79	1.53	0.00
Developed professional identity	8	8.07	0.47	0.00
Problem-solving ability	8	8.00	0.88	0.50
Knowledge of contemporary issues	8	8.07	0.73	0.00
Respect for oneself	8	7.93	0.73	0.00
Knowledge of safety-procedures	8	8.07	1.38	0.00
Working in multidisciplinary teams	8	8.00	0.88	0.50
Knowledge of engineering designs	8	8.07	1.00	0.00
Knowledge of quality control measures	8	7.86	1.61	0.00
Site-supervision knowledge	8	7.71	1.44	0.50
Financial management skill	8	7.71	1.20	0.00
Interpersonal relationship with industry professionals	8	7.71	1.44	0.00
Networking with other interns on-site	8	7.86	0.95	0.00
Exposure to career opportunities	9	8.57	1.34	0.00
Enhanced learning	9	8.79	1.05	0.00
Acquisition of professional skills	9	8.50	1.83	0.00
Technical report writing	8	8.07	1.14	0.00
Mentorship opportunities	8	8.07	1.33	0.75

value of 9, indicating very good consensus. This implies that these factors have the greatest significance regarding graduate employability. Apart from this, a total of 21 other factors possessed a strong median value of 8 which indicates a high significance regarding graduate employability. Consensus was further measured by inspecting IQD values, as seen in Table 8.13. As previously stated, factors with IQD value less than 1 were considered to have high agreement and consensus among experts. All 28 work-integrated learning factors satisfied these criteria, hence consensus was fully achieved after the second round.

Emotional intelligence factors (first and second round Delphi results)

During the first round of the Delphi process, a total of 18 emotional intelligence factors were identified from the literature review, as shown in Table 8.14. Experts were required to indicate the significance of each factor in contributing to graduate employability. The rating was achieved using a 10-point Likert scale of 'no significance', 'low significance', 'medium significance', 'high significance', and 'very high significance'. While 'no significance' was assigned the lowest weighting (1 and 2), 'very high significance' had the highest weighting (9 and 10). At this stage, IQD above 1.00 were recorded which necessitated a second round.

No new additions were made during the first round and consensus was achieved after the second round, as shown in Table 8.15.

Table 8.14 Emotional intelligence factors (first round results)

Emotional intelligence factors	Median	Mean ($\bar{x}$)	SD (σX)	IQD
Perceive and accurately express emotion	8	7.94	1.98	3.00
Understand and manage emotions	9	8.41	1.50	2.00
Improved decision-making	9	8.53	0.87	1.00
Self-awareness and self-control	8	8.12	1.11	1.00
Improved self-motivation	8	7.94	1.25	2.00
Communicate effectively	8	8.12	1.41	2.00
Relationship management	8	7.59	1.87	2.00
Empathy and compassion for others	8	7.29	2.42	2.00
Emotional reasoning	8	7.35	2.23	3.00
Increased confidence and self-management	8	8.00	1.50	1.00
Increased social awareness	8	7.35	1.80	3.00
Positive outlook to life	8	7.65	1.93	3.00
Flexibility in solving problems	9	8.47	1.66	2.00
Building effective relationships	8	7.71	1.93	2.00
Conflict management skills	8	7.82	1.94	2.00
Ability to adapt to new ideas and surroundings	9	8.29	1.61	2.00
Manage stress	9	8.35	1.93	2.00
Ability to display resilience	9	8.24	1.92	1.00

Table 8.15 Emotional intelligence factors (second round results)

Emotional intelligence factors	Median	Mean ($\bar{x}$)	SD (σX)	IQD
Perceive and accurately express emotion	8	7.79	0.89	0.00
Understand and manage emotions	9	8.71	0.91	0.00
Improved decision-making	9	8.79	0.58	0.00
Self-awareness and self-control	8	8.07	0.47	0.00
Improved self-motivation	8	7.93	0.47	0.00
Communicate effectively	8	7.86	0.86	0.00
Relationship management	8	7.79	1.19	0.00
Empathy and compassion for others	8	7.71	1.44	0.00
Emotional reasoning	8	7.71	1.49	0.00
Increased confidence and self-management	8	7.93	0.73	0.00
Increased social awareness	8	7.79	1.05	0.00
Positive outlook to life	8	7.86	1.17	0.00
Flexibility in solving problems	9	8.79	1.12	0.00
Building effective relationships	8	7.64	1.39	0.00
Conflict management skills	8	7.86	1.10	0.00
Ability to adapt to new ideas and surroundings	9	8.57	1.16	0.00
Manage stress	9	8.57	1.16	0.00
Ability to display resilience	9	8.57	1.34	0.00

As seen from Table 8.15, *understand and manage emotions, improved decision-making, flexibility in solving problems, ability to adapt to new ideas and surroundings, manage stress,* and *ability to display resilience in the face of adversity* all possess a strong median value of 9, indicating very good consensus. This implies that these factors have the greatest significance for graduate employability. Aside from this, a total of 12 other factors possessed a strong median value of 8 which indicates a high significance for graduate employability. Consensus was further measured by inspecting IQD values, as seen in Table 8.15. As previously stated, factors with IQD value less than 1 were considered to have high agreement and consensus among experts. All 18 emotional intelligence factors satisfied these criteria, hence consensus was fully achieved after the second round.

University–industry collaboration (UIC) (first and second round Delphi results)

During the first round of the Delphi process, a total of 16 UIC factors were identified from the literature review, as shown in Table 8.16. Experts were required to indicate the significance of each factor in contributing to graduate employability. The rating was achieved using a 10-point Likert scale of 'no significance', 'low significance', 'medium significance', 'high significance', and 'very high significance'. While 'no significance' was assigned the lowest weighting (1 and 2), 'very high significance' had the highest weighting (9 and 10). At this stage, IQD above 1.00 were recorded which necessitated a second round.

Table 8.16 University–industry collaboration factors (first round results)

UIC factors	Median	Mean ($\bar{x}$)	SD (σX)	IQD
Scholarship opportunities for students	9	8.47	1.70	3.00
Job opportunities for graduates	9	8.59	1.28	2.00
Number of research grants from industry	9	8.00	1.70	2.00
Existence of open-day events	7	7.65	1.73	2.00
Field trips to industry	8	8.00	1.54	2.00
Industry mentoring for students	8	8.12	1.50	2.00
Vocational training	8	8.12	1.45	2.00
Invited guest speakers from industry	8	7.88	1.65	3.00
Networking with industry professionals	8	8.06	1.82	3.00
Entrepreneurial culture in universities	8	7.88	2.00	2.00
Publication opportunities with industry	8	7.41	2.09	2.00
Construction project exercises in classrooms	9	8.18	1.63	2.00
Patenting opportunities	9	7.81	2.07	2.25
Training programmes opportunities for students	8	8.18	1.70	1.00
Student's exposure to innovative ideas	9	8.43	1.40	0.00
Involvement in construction projects	8	7.79	0.80	0.00

Table 8.17 University–industry collaboration factors (second round results)

UIC factors	Median	Mean ($\bar{x}$)	SD (σX)	IQD
Scholarship opportunities for students	9	8.93	0.73	0.00
Job opportunities for graduates	9	8.79	0.80	0.00
Number of research grants from industry	9	8.43	1.50	0.75
Existence of open-day events	7	6.71	0.83	0.75
Field trips to industry	8	7.64	1.01	0.00
Industry mentoring for students	8	7.79	0.43	0.00
Vocational training	8	7.57	0.94	0.00
Invited guest speakers from industry	8	7.71	0.99	0.00
Networking with industry professionals	9	8.50	1.56	0.00
Entrepreneurial culture in universities	8	7.79	1.37	0.00
Publication opportunities with industry	8	7.43	1.40	0.00
Construction project exercises in classrooms	9	8.29	1.33	0.75
Patenting opportunities	9	8.64	1.08	0.00
Training programmes opportunities for students	9	8.50	1.40	0.00
Student's exposure to innovative ideas	9	8.43	1.40	0.00
Involvement in construction projects	8	7.79	0.80	0.00

No new additions were made during the first round and consensus was achieved after the second round, as shown in Table 8.17.

As seen in Table 8.17, eight of the factors possess a strong median value of 9, indicating a very good consensus. This implies that these factors have the greatest significance for graduate employability. Aside from this, a total of seven other factors possessed a strong median value of 8 which indicates a high significance for graduate employability. With a median value of 7 and mean value of 6.71, *existence of open-day events* was the least ranked, but still ranks under high significance in the measurement scale. Consensus was further measured by inspecting IQD values, as seen in Table 8.17. As previously stated, factors with IQD value less than 1 were considered to have high agreement and consensus among experts. All 16 UIC factors satisfied these criteria, hence consensus was fully achieved after the second round.

Industry 4.0 requirements (4IR) (first and second round Delphi results)

During the first round of the Delphi process, a total of 20 4IR requirements were identified from the literature review, as shown in Table 8.18. Experts were required to indicate the significance of each skill in contributing to graduate employability. The rating was achieved using a 10-point Likert scale of 'no significance', 'low significance', 'medium significance', 'high significance', and 'very high significance'. While 'no significance' was assigned the lowest weighting (1 and 2), 'very high significance' had the highest weighting (9 and 10).

However, one of the experts suggested three additional factors: *cryptocurrency*, *mechatronics*, and *information technology and systems control*. All three were included

Table 8.18 Industry 4.0 requirements (first round results)

4IR requirements	Median	Mean ($\bar{x}$)	SD (σX)	IQD
Knowledge of the use of drones	8	6.53	3.12	5.00
Internet of Things (use of Internet)	8	7.59	2.45	2.00
Building information modelling (BIM)	8	7.71	1.90	2.00
Robotics knowledge	8	6.94	2.95	3.00
Internet of Services	8	7.71	2.37	1.00
Modelling-based tools	8	7.47	2.10	3.00
Advanced data analytics	8	7.65	2.32	2.00
Virtual reality	8	6.88	2.57	2.00
Augmented reality	8	6.94	2.61	3.00
Mixed reality	8	7.00	2.42	2.00
Knowledge of cybersecurity	8	7.29	2.20	2.00
Knowledge of mobile devices	8	7.47	2.12	2.00
Knowledge of artificial intelligence	8	7.47	2.35	3.00
Knowledge of blockchain/bitcoin	8	6.82	3.05	3.00
Engineering design (computer-aided design)	8	8.06	1.56	1.00
Electronics technology	8	7.06	2.16	1.00
Self-service use of E-library	8	7.53	1.81	2.00
Cloud computing	8	7.47	2.27	3.00
Digital media (Internet, social media, etc.)	8	7.18	2.48	4.00
Administrative tools	8	8.41	1.54	2.00

Table 8.19 Industry 4.0 requirements (second round results)

4IR requirements	Median	Mean ($\bar{x}$)	SD (σX)	IQD
Knowledge of the use of drones	8	7.21	1.81	0.00
Internet of Things (use of Internet)	8	7.93	1.07	0.75
Building information modelling (BIM)	8	8.00	0.88	0.00
Robotics knowledge	8	7.50	2.03	0.00
Internet of Services	8	8.00	1.30	0.00
Modelling-based tools	8	7.64	1.50	0.00
Advanced data analytics	8	7.71	1.77	0.00
Virtual reality	8	7.50	1.56	0.00
Augmented reality	8	7.64	1.50	0.75
Mixed reality	8	7.64	1.39	0.00
Knowledge of cybersecurity	8	7.50	1.74	0.00
Knowledge of mobile devices	8	7.50	1.74	0.00
Knowledge of artificial intelligence	8	7.50	1.74	0.00
Knowledge of blockchain/bitcoin	8	7.36	2.13	0.00
Engineering design (computer-aided design)	8	8.00	1.04	0.00
Electronics technology	8	7.29	1.86	0.75
Self-service use of E-library	8	7.57	1.09	0.00
Cloud computing	8	7.50	1.87	0.00
Digital media (Internet, social media, etc.)	8	7.29	2.05	0.75
Administrative tools	8	7.79	1.12	0.00
Cryptocurrency	8	7.71	1.38	0.00
Mechatronics	8	7.57	1.40	0.75
Information technology and systems control	9	8.86	0.66	0.75

in the second round and experts were requested to rank their level of significance. After the second round, consensus was achieved, as shown in Table 8.19.

As seen in Table 8.19, *information technology and systems control* possesses a strong median value of 9, indicating very good consensus. This implies that these factors have the greatest significance for graduate employability. Apart from this, all other factors possessed a strong median value of 8 which indicates a high significance for graduate employability. Consensus was further measured by inspecting IQD values, as seen in Table 8.19. As previously stated, factors with IQD value less than 1 were considered to have high agreement and consensus among experts. All 23 4IR requirements satisfied these criteria, hence a consensus was fully achieved after the second round.

Discussion of Delphi study results

Following the presentation of the findings from the Delphi study, this section of the book discusses and validates the various dimensions and constructs established by the panel of experts or perceived as significant and having an influence on improved graduate employability of built environment graduates.

This study identified six key constructs that contributes to the employability framework for built environment graduates in the 4IR era. One of the critical dimensions that contribute to the employability of students is the possession of generic skills. Considering the various dynamics and complexities of the industry, including increased global competition, technological advancements, environmental factors, and construction projects sophistication, the need for graduates to possess generic skills cannot be overstated. These changes require graduates to possess not only a sound academic background but also non-academic skills and competencies. The experts for this study identified generic skills as a key dimension with an absolute median score of 10, highlighting their agreement on its importance for graduate employability. The consensus reached by experts corroborates the findings from literature which reinforces the need for graduates to possess these non-academic skills. Generic skills such as communication skills, teamwork skills, problem-solving skills, adaptability skills, critical thinking (creative thinking) skills, organizing and management skills, lifelong learning skills, and interpersonal skills have been found to improve graduate employability in the eyes of employers (Love *et al.*, 2002; Gardner *et al.*, 2005; Washer, 2007; Kilgour and Koslow, 2009; Farooqui and Ahmed, 2009; Ariana, 2010; Wickramasinghe and Perera, 2010; Rawlins and Marasini, 2011; Ahn *et al.*, 2012; Ayarkwa *et al.*, 2012; Lievens and Sackett, 2012; Reid and Anderson, 2012; Jackson and Chapman, 2012; Samavedham and Ragupathi, 2012; Finch *et al.*, 2013; Aliu and Aigbavboa, 2017; O'Leary, 2017; Nguyen, 2019). Other generic skills considered as important by experts and confirmed by literature include information and technology (ICT) skills, entrepreneurship skills, leadership skills, and ethical skills (Russell *et al.*, 2007; Tatum, 2010; Arain, 2010; Ahn *et al.*, 2012; Conrad and Newberry, 2012; Durrani and Tariq, 2012; Nowiński *et al.*, 2019). With a strong median value of 9, experts reached a strong consensus in critical thinking skills, creativity/innovation skills, willingness to learn, information and communication technology (ICT) skills, and problem-solving skills, meaning these five sets of skills have the greatest significance on graduate employability. This resonates with the studies of Wickramasinghe and Perera (2010), Rawlins and Marasini (2011), Ahn *et al.* (2012), Ayarkwa *et al.* (2012), Lievens and Sackett (2012), Reid and Anderson (2012), Jackson and Chapman (2012), Samavedham and Ragupathi (2012), Finch *et al.* (2013), Aliu and Aigbavboa (2017), and O'Leary (2017). The other skills were ranked highly with median values of 8, implying that they have an equal significance on graduate employability as confirmed from the literature. Moreover, the generic skills under consideration all had an IQD value of less than 1.00, hence consensus was fully achieved.

Another key dimension identified is the possession of discipline-specific skills that students have to possess during their academic cycle. The experts for this study identified discipline-specific skills as a key dimension with a median score of 9, highlighting their agreement on its importance for graduate employability. Some of these knowledge and skills acquired by virtue of studying an academic discipline include knowledge of subject area and curricula, critical thinking pedagogical skills, organizational skills, communication skills, administrative skills, management skills, lifelong learning, decision-making skill, digital literacy, self-confidence, connection with other disciplines, problem-solving skills, collaborative skills, transformative skills, and hands-on experience. This set of skills confirms the works of Dacre Pool and Sewell (2007), Rae (2007), Bridgstock (2009), Coetzee and Beukes (2010), Hutchinson (2013), Watts (2013), Nagarajan and Edwards (2015), and Lamanauskas and Augienė (2017). With a strong median value of 10 and mean values of 9.33 and 9.44, respectively, experts reached a strong consensus in the knowledge of the subject area and lifelong learning, indicating that both sets of skills have the greatest significance for graduate employability. This resonates with the studies of Rae (2007), Bridgstock (2009), Coetzee and Beukes (2010), and Hutchinson (2013). Moreover, the discipline-specific skills under consideration all had an IQD value of less than 1.00, hence consensus was fully achieved.

The importance of work experience in improving the employability of built environment graduates cannot be quantified, hence its inclusion in this study. By engaging in work activities, students are presented with enhanced knowledge about their career choices and related occupations. The experts for this study identified work experience as a key dimension with a median score of 9, highlighting their agreement on its importance on graduate employability. Some of these knowledge and skills acquired by undergoing work experience (work-integrated learning [WIL]) include developing practical knowledge, understanding job responsibilities, awareness of industry, improved work attitude, developing interpersonal values, résumé development, awareness of workplace culture, time management, self-confidence, problem-solving ability, working in multidisciplinary teams, site-supervision knowledge, financial management skills, interpersonal relationships with industry professionals, networking with other interns on-site, exposure to career opportunities, acquisition of professional skills, technical report writing, and mentorship opportunities. This set of skills confirms the works of Gault *et al.* (2000), Gill and Lashine (2003), Mihail (2006), Wasonga and Murphy (2006), Divine *et al.* (2007), McLennan and Keating (2008), Omar *et al.* (2008), Cooper *et al.* (2010), Sattler *et al.* (2011), Lowden *et al.* (2011), and Pillai *et al.* (2012). With a strong median value of 9, experts reached a strong consensus in time management, developing practical knowledge, understand job responsibilities, awareness of the industry, exposure to career opportunities, enhanced learning, and acquisition of professional skills, indicating that these factors have the greatest significance on graduate employability. This resonates with the studies of Sattler *et al.* (2011), Lowden *et al.* (2011), and Pillai *et al.* (2012). Moreover, the work experience factors under consideration all had an IQD value of less than 1.00, hence consensus was fully achieved.

Furthermore, emotional intelligence was considered a key dimension for graduate employability. Various studies state that individuals with high levels of EI tend to motivate others and themselves which can lead to successful careers. The experts for this study identified emotional intelligence as a key dimension with a median score of 9, highlighting their agreement on its importance for graduate employability. Individuals who exhibit high levels of emotional intelligence have been found to perceive and accurately express emotion, make decisions effectively, communicate effectively, manage relationships, show compassion for others, solve problems flexibly, build effective relationships, adapt to new ideas and surroundings, manage stress and display resilience in the face of adversity. This set of skills confirms the works of Goleman (1998), Cherniss (2000), Fugate *et al.* (2004), Lopes *et al.* (2006), Blickle and Witzki (2008), Coetzee and Beukes (2010), Stein and Book (2011), and Jameson *et al.* (2016). With a strong median value of 9, experts reached a strong consensus in skills such as understanding and managing emotions, improved decision-making, flexibility in solving problems, ability to adapt to new ideas and surroundings, managing stress, and ability to display resilience in the face of adversity, meaning these factors have the greatest significance for graduate employability. This resonates with the studies of Blickle and Witzki (2008), Coetzee and Beukes (2010), Stein and Book (2011), and Jameson *et al.* (2016). Moreover, the emotional intelligence factors under consideration all had an IQD value of less than 1.00, hence consensus was fully achieved.

Furthermore, in enhancing the employability of graduates for the industry, the collaboration between universities and industry plays a key role and its importance cannot be ignored. With the present-day industry driven by various factors, including technology and globalization, it is imperative for universities to establish cordial connections with industry counterparts in a bid to provide opportunities for students to learn more about these significant changes. The experts for this study identified university–industry collaborations (UIC) as a key dimension with a median score of 9, highlighting their agreement on its importance for graduate employability. Several roles of UIC exist in developing graduate employability, including enhancing job opportunities for graduates, encouragements of field trips to industry, industry mentoring for students, vocational training, networking activities with industry professionals, publication opportunities with industry, patenting opportunities, training programme opportunities for students, students' exposure to innovative ideas and involvement in construction projects. These roles confirm the works of Hagan (2004), Feng *et al.* 2011), Ramakrishnan and Yasin (2011), Haupt (2012), and Turhan and Akman (2013). Experts reached a strong consensus in eight of the factors with a strong median value of 9, indicating that these factors have the greatest significance on graduate employability. This amplifies the studies of Haupt (2012) and Turhan and Akman (2013). More so, the UIC factors under consideration all had an IQD value of less than 1.00, hence consensus was fully achieved.

Further findings highlighted the need for built environment graduates to possess certain skills and requirements in the era of the 4IR. With the advent of high-speed networks and smart infrastructures dominating our present-day world, there is a new degree of transformation, not experienced since the first industrial revolution. In embracing the fourth industrial revolution, built environment graduates will require a plethora of skills as automation continues to advance. The experts for this study identified 4IR skills as a key dimension with a median score of 10, highlighting their agreement on its importance for graduate employability. Some of these skills and attributes include the ability to determine the deeper meaning of innovations (sense-making skills), ability to connect to others deeply and directly (social intelligence), ability to operate and adapt to different cultural settings (cross-cultural competence), proficiency in proffering solutions (adaptive thinking), ability to understand data-based reasoning (computational thinking), ability to assess and develop content in advanced media forms (new-media literacy), ability to understand concepts across numerous disciplines (transdisciplinarity), and ability to develop diverse tasks and workflows for positive outcomes (design mindset). These skills and abilities confirm the works of Madsen *et al.* (2016), Benesova *et al.* (2017), Cotet *et al.* (2017), Grzybowska and Łupicka (2017), Pinzone *et al.* (2017), Ras *et al.* (2017), Caldarola *et al.* (2018), and Schallock *et al.* (2018). With a strong median value of 9, experts reached strong consensus in information technology and systems control, indicating that these factors have the greatest significance for graduate employability. This resonates with the studies of Caldarola *et al.* (2018) and Schallock *et al.* (2018). Moreover, 4IR factors under consideration all had an IQD value of less than 1.00, hence consensus was fully achieved.

In conclusion, therefore, the various dimensions (main and sub-dimensions) that contribute to the improved employability of built environment graduates in South Africa were validated by the experts. Hence, the experts for this study were in full agreement with the adoption of these six dimensions in developing an employability improvement framework for built environment graduates.

Summary

This chapter focuses on the validation of the constructs of the proposed employability framework by adopting a Delphi approach using South Africa as a case study. This chapter presented a summary of the first and second round Delphi results as well as subsequent discussions. The chapter also presented the computation for every construct to determine their significance in developing an employability improvement framework for built environment graduates in South Africa. Based on the Delphi-specific objectives, this chapter also discussed the findings and supported them from literature. Most importantly, a consensus was reached by experts for each construct, which improved the reliability of this study. The Delphi results played a significant role in determining the various dimensions (main and sub-dimensions) that contribute to the improved employability of built environment graduates in South Africa. These constructs led to the development of an employability improvement framework for built environment graduates in South Africa.

References

Adler, M. and Ziglio, E. 1996. *Gazing into the Oracle: The Delphi method and its application to social policy and public health.* London: Kingsley Publishers.
Ahn, Y.H., Annie, R.P. and Kwon, H. 2012. Key competencies for US construction graduates: Industry perspective. *Journal of Professional Issues in Engineering Education and Practice, 138*(2), pp. 123–130. doi: 10.1061/(ASCE)EI.1943-55410000089.

Aliu, O.J. and Aigbavboa, C. 2017. *Upscaling construction education to meet the needs of the Nigerian construction industry* (Master's dissertation, University of Johannesburg, Johannesburg).

Andranovich, G. 1995. *Developing community participation and consensus: The Delphi technique.* Pullman, WA: USDA.

Arain, F.M. 2010. Identifying competencies for baccalaureate level construction education: Enhancing employability of young professionals in the construction industry. In *Proceedings of the Construction Research Congress 2010: Innovation for Reshaping Construction Practice*, Construction Research Congress 2010, May 8–10, 2010, American Society of Civil Engineers (ASCE), Banff, AB, pp. 194–204.

Archer, W. and Davison, J. 2008. *Graduate employability: What employers want?* London: The Council for Industry and Higher Education.

Ariana, S.M. 2010. Some thoughts on writing skills. *Annals of the University of Oradea, Economic Science Series*, 19(1), pp. 134–140.

Ayarkwa, J., Adinyira, E. and Osei-Asibey, D. 2012. Industrial training of construction students: Perceptions of training organizations in Ghana. *Education Training*, 54(2/3), pp. 234–249. http://dx.doi.org/10.1108/00400911211210323.

Benesova, A. and Tupa, J. 2017. Requirements for education and qualification of people in Industry 4.0. *Procedia Manufacturing*, 11, pp. 2195–2202.

Blickle, G. and Witzki, A. 2008. New psychological contracts in the world of work: Economic citizens or victims of the market? *Society and Business Review*, 3(2), pp. 149–161.

Bridgstock, R. 2009. The graduate attributes we've overlooked: Enhancing graduate employability through career management skills. *Higher Education Research & Development*, 28(1), pp. 31–44. doi: 10.1080/07294360802444347.

Buckley, C. 1995. Delphi: A methodology for preferences more than predictions. *Library Management*, 16(7), pp. 16–19.

Caldarola, E.G., Modoni, G.E. and Sacco, M. 2018. A knowledge-based approach to enhance the workforce skills and competences within the industry 4.0. *Proceedings of the Tenth International Conference on Information, Process, and Knowledge Management*, Rome.

Cavalli-Sforza, V. and Ortolano, L. 1984. Delphi forecasts of land use–transportation interactions. *Journal of Transportation Engineering*, 110(3), pp. 324–339.

Chan, A.P., Yung, E.H., Lam, P.T., Tam, C.M. and Cheung, S. 2001. Application of Delphi method in selection of procurement systems for construction projects. *Construction Management and Economics*, 19(7), pp. 699–718.

Cherniss, C. 2000. Emotional intelligence: What it is and why it matters. *Paper presented at the annual meeting of the Society for Industrial and Organizational Psychology*, New Orleans, USA.

Coetzee, M. and Beukes, C.J. 2010. Employability, emotional intelligence and career preparation support satisfaction among adolescents in the school-to-work transition phase. *Journal of Psychology in Africa*, 20(3), pp. 439–446.

Conrad, D. and Newberry, R. 2012. Identification and instruction of important business communication skills for graduate business education. *Journal of Education for Business*, 87(2), pp. 112–120.

Cooper, L., Orrell, J. and Bowden, M. 2010. *Work integrated learning: A guide to effective practice.* New York: Routledge.

Cotet, G.B., Balgiu, B.A. and Zaleschi, V.C. 2017. Assessment procedure for the soft skills requested by industry 4.0. *MATEC Web of Conferences* (Vol. 121, p. 07005). Bucharest: EDP Sciences.

Creswell, J.W. 2014. *Research design: Qualitative, quantitative and mixed methods approaches* (4th ed.). Lincoln: Sage Publications.

Critcher, C. and Gladstone, B. 1998. Utilizing the Delphi technique in policy discussion: A case study of a privatized utility in Britain. *Public Administration, 76*(3), pp. 431–449.

Dacre Pool, L. and Sewell, P. 2007. The key to employability: Developing a practical model of graduate employability. *Education Training, 49*(4), pp. 277–289.

Dalkey, N.C. and Helmer, O. 1963. An experimental application of the Delphi method to the use of experts. *Journal of the Institute of Management Sciences*, 9, pp. 458–467.

Delbecq, A.L., Van de Ven, A.H. and Gustafson, D.H. 1975. *Group techniques for program planning: A guide to nominal group and Delphi processes.* Glenview, IL: Scott, Foresman & Company.

Divine, R.L., Linrud, J.K., Miller, R.H. and Wilson, J.H. 2007. Required internship programs in marketing: Benefits, challenges and determinants of fit. *Marketing Education Review, 17*(2), pp. 45–52. http://dx.doi.org/10.1080/10528008.2007.11489003.

Durrani, N. and Tariq, V.N. 2012. The role of numeracy skills in graduate employability. *Education Training, 54*(5), pp. 419–434. http://dx.doi.org/10.1108/00400911211244704.

Farooqui, R.U. and Ahmed, S.M. 2009. Key skills for graduating construction management students: A comparative study of industry and academic perspectives. *Proceedings of the Construction Research Congress 2009: Building a Sustainable Future*, American Society of Civil Engineers (ASCE), Washington, DC, p. 1439.

Feng, C., Ding, M. and Sun, B. 2011. A comparison research on industry-university-research strategic alliances in countries. *Asian Social Science, 7*(1), p. 102.

Finch, D.J., Hamilton, L.K., Baldwin, R. and Zehner, M. 2013. An exploratory study of factors affecting undergraduate employability. *Education Training, 55*(7), pp. 681–704. http://dx.doi.org/10.1108/ET-07-2012-0077.

Fugate, M. and Ashforth, B. 2004. Employability: The construct, its dimensions, and applications. *Proceedings of the Annual Meeting of the Academy of Management, OB: J1-J6*, Seattle.

Gardner, C.T., Milne, M.J., Stringer, C.P. and Whiting, R.H. 2005. Oral and written communication apprehension in accounting students: Curriculum impacts and impacts on academic performance. *Accounting Education, 14*(3), pp. 313–336.

Gault, J., Redington, J. and Schlager, T. 2000. Undergraduate business internships and career success: Are they related? *Journal of Marketing Education, 22*(1), pp. 45–53.

Gill, A. and Lashine, S. 2003. Business education: A strategic market-oriented focus. *International Journal of Educational Management, 175*, pp. 188–194. http://dx.doi.org/10.1108/09513540310484904.

Goleman, D. 1998. *Working with emotional intelligence.* New York: Bantam.

Green, B., Jones, M., Hughes, D. and Williams, A. 1999. Applying the Delphi technique in a study of GP's information requirements. *Health and Social Care in the Community, 17*(3), pp. 198–205.

Grisham, T. 2008. *Cross-cultural leadership* (Doctor of Project Management, School of Property, Construction and Project Management, RMIT, Melbourne).

Grzybowska, K. and Łupicka, A. 2017. Key competencies for industry 4.0. *Economics & Management Innovations, 1*(1), pp. 250–253.

Häder, M. and Häder, S. 1995. Delphi and cognitive psychology: An approach to the theoretical foundation of the Delphi method. *ZUMA News, 19*(37), pp. 8–34.

Hagan, D. 2004. Employer satisfaction with ICT graduates. *Proceedings of the Sixth Australasian Conference on Computing Education*, Australian Computer Society, Inc. Sydney, Vol. 30, pp. 119–123.

Hallowell, M. and Gambatese, J. 2010. Qualitative research: Application of the Delphi method to CEM research. *Journal of Construction Engineering and Management, 136*(Special Issue: Research Methodologies in Construction Engineering and Management), pp. 99–107.

Hasson, F., Keeney, S. and McKenna, H. 2000. Research guidelines for the Delphi survey technique. *Journal of Advanced Nursing, 32*(4), pp. 1008–1015.

Haupt, T. 2012. Student attitudes towards cooperative construction education experiences. *Construction Economics and Building, 3*(1), pp. 31–42.

Holey, E.A., Feeley, J.L., Dixon, J. and Whittaker, V.J. 2007. An exploration of the use of simple statistics to measure consensus and stability in Delphi studies. BMC *Medical Research Methodology, 7*(52), pp. 1–10.

Hsu, C.C. and Sandford, B.A. 2007. The Delphi technique: Making sense of consensus. *Practical Assessment, Research and Evaluation, 12*(10), pp. 1–8.

Hutchinson, J. 2013. *School organisation and STEM career-related learning.* Derby: International Centre for Guidance Studies. Retrieved from: http://derby.openrepository.com/derby/bitstream/10545/303288/8/STEM%20Leaders%20Report%202013%20%28High%20res%29.pdf.

Jackson, D. and Chapman, E. 2012. Non-technical competencies in undergraduate business degree programs: Australian and UK perspectives. *Studies in Higher Education, 37*(5), pp. 541–567. doi: 10.1080/03075079.2010.527935.

Jameson, A., Carthy, A., McGuinness, C. and McSweeney, F. 2016. Emotional intelligence and graduates – employers' perspectives. *Procedia – Social and Behavioural Sciences, 228,* pp. 515–522.

Kilgour, M. and Koslow, S. 2009. Why and how do creative thinking techniques work? Trading off originality and appropriateness to make more creative advertising. *Journal of the Academy of Marketing Science, 37*(3), pp. 298–309.

Lam, S.S.Y., Petri, K.L. and Smith, A.E. 2000. Prediction and optimization of a ceramic casting process using a hierarchical hybrid system of neural networks and fuzzy logic. *IIE Transactions, 32*(1), pp. 83–92.

Lamanauskas, V. and Augienė, D. 2017. Career education in comprehensive schools of Lithuania: Career professionals' position. In A. Klim-Klimaszewska, M. Podhajecka and A. Fijalkowska-Mroczek (Eds.), *Orientacje i przedsięwzięcia w edukacjiprzedszkolnej i szkolnej /Monografia/* (pp. 216–236). Siedlce: Akka.

Lievens, F. and Sackett, P.R. 2012. The validity of interpersonal skills assessment via situational judgment tests for predicting academic success and job performance. *Journal of Applied Psychology, 97*(2), p. 460.

Linstone, A. and Turoff, M. 1978. *The Delphi method: Techniques and applications.* Reading, MA: Addison-Wesley.

Linstone, H.A. and Turoff, M. 2002. *The Delphi method: Techniques and applications.* Retrieved from: www.is.njit.edu/pubs.php [Accessed 12 December 2011].

Loo, R. 2002. The Delphi method: A powerful tool for strategic management, policing. *International Journal of Police Strategies and Management, 25*(4), pp. 762–769.

Lopes, P.N., Grewal, D., Kadis, J., Gall, M. and Salovey, P. 2006. Evidence that emotional intelligence is related to job performance and affect and attitudes at work. *Psichothemia, 18*(1), pp. 132–138.

Love, P., Haynes, N., Sohal, A., Chan, A. and Tam, C. 2002. *Key construction management skills for future success.* Monash University. Faculty of Business and Economics. Working Paper 49/02.

Lowden, K., Hall, S., Ellio, D.D. and Lewin, J. 2011. *Employers' perceptions of the employability skills of new graduates*. London: Edge Foundation.

Madsen, E.S., Bilberg, A. and Hansen, D.G. 2016. Industry 4.0 and digitalization call for vocational skills, applied industrial engineering, and less for pure academics. *Proceedings of the 5th P&OM World Conference, Production and Operations Management, P&OM,* Havana International Conference Center, Havana.

McKenna, H. 1994. The Delphi technique: A worthwhile research approach for nursing? *Journal of Advanced Nursing, 19*(6), pp. 1221–1225.

McLennan, B. and Keating, S. 2008, June. Work-integrated learning (WIL) in Australian universities: The challenges of mainstreaming WIL. *Proceedings of the ALTC NAGCAS National Symposium*, Melbourne, Australia, pp. 2–14.

Mihail, D.M. 2006. Internships at Greek universities: An exploratory study. *Journal of Workplace Learning, 18*(1), pp. 28–41. doi: 10.1108/13665620610641292.

Nagarajan, S. and Edwards, J. 2015. The role of universities, employers, graduates and professional associations in the development of professional skills of new graduates. *Journal of Perspectives in Applied Academic Practice, 3*(2).

Nguyen, P. 2019. *Enhancing the employability of graduate students with transversal skills* (Bachelor's thesis, Lahti University of Applied Sciences, Lahti).

Nowiński, W., Haddoud, M.Y., Lančarič, D., Egerová, D. and Czeglédi, C. 2019. The impact of entrepreneurship education, entrepreneurial self-efficacy and gender on entrepreneurial intentions of university students in the Visegrad countries. *Studies in Higher Education, 44*(2), pp. 361–379.

Okoli, C. and Pawlowski, S.D. 2004. The Delphi method as a research tool: An example, design considerations and applications. *Information & Management, 42*(1), pp. 15–29.

O'Leary, S. 2017. Graduates' experiences of, and attitudes towards, the inclusion of employability-related support in undergraduate degree programmes; Trends and variations by subject discipline and gender. *Journal of Education and Work, 30*(1), pp. 84–105.

Omar, M.Z., Rahman, M.N.A., Kofli, N.T., Mat, K., Darus, M.Z. and Osman, S.A. 2008. Assessment of engineering students' perception after industrial training placement. *Proceedings of 4th WSEAS/IASME International Conference on Educational Technologies (EDUTE'08)*, Corfu, Greece.

Phillips, R. 2000. *New applications for the Delphi technique*. San Diego, CA: Pfeiffer and Company.

Pillai, S., Khan, M.H., Ibrahim, I.S. and Raphael, S. 2012. Enhancing employability through industrial training in the Malaysian context. *Higher Education, 63*(2), pp. 187–204. doi: 10.1007/s10734-011-9430-2.

Pinzone, M., Fantini, P., Perini, S., Garavaglia, S., Taisch, M. and Miragliotta, G. 2017, September. Jobs and skills in industry 4.0: An exploratory research. *Proceedings of the IFIP International Conference on Advances in Production Management Systems*, Springer, Cham, pp. 282–288.

Rae, D. 2007. Connecting enterprise and graduate employability: Challenges to the higher education culture and curriculum? *Education+ Training, 49*(8/9), pp. 605–619.

Ramakrishnan, K. and Yasin, N.M. 2011. Higher learning institution-industry collaboration: A necessity to improve teaching and learning process. *Proceedings of the Computer Science & Education (ICCSE) 2011 6th International Conference on IEEE*, Singapore, p. 1445.

Ras, E., Wild, F., Stahl, C. and Baudet, A. 2017, June. Bridging the skills gap of workers in industry 4.0 by human performance augmentation tools: Challenges and roadmap.

Proceedings of the 10th International Conference on Pervasive Technologies Related to Assistive Environments, Island of Rhodes, Greece, pp. 428–432.

Raskin, M.S. 1994. The Delphi study in field instruction revisited: Expert consensus on issues and research priorities. *Journal of Social Work Education*, 30, pp. 75–89.

Rawlins, J. and Marasini, R. 2011. Are the construction graduates on CIOB accredited degree courses meeting the skills required by the industry? *Management*, 167, p. 174.

Rayens, M.K. and Hahn, E.J. 2000. Building consensus using the policy Delphi method. *Policy Politics Nursing Practice*, 1(2), pp. 308–315.

Reid, J.R. and Anderson, P.R. 2012. Critical thinking in the business classroom. *Journal of Education for Business*, 87(1), pp. 52–59.

Rogers, M.R. and Lopez, E.C. 2002. Identifying critical cross-cultural school psychology competencies. *Journal of School Psychology*, 40(2), pp. 115–141.

Rowe, G., Wright, G. and Bolger, F. 1991. Delphi: A re-evaluation of research and theory. *Technological Forecasting and Social Change*, 39, pp. 238–251.

Russell, J.S., Hanna, A., Bank, L.C. and Shapira, A. 2007. Education in construction engineering and management built on tradition: Blueprint for tomorrow. *Journal of Construction Engineering and Management*, 133(9), pp. 661–668. doi: 10.1061/ (ASCE)0733-9364(2007)133:9(661).

Samavedham, L. and Ragupathi, K. 2012. Facilitating 21st century skills in engineering students. *Journal of Engineering Education Transformations*, 25(4–1), pp. 37–49.

Sattler, P., Wiggers, R.D. and Arnold, C. 2011. Combining workplace training with post-secondary education: The spectrum of work-integrated earning (WIL) opportunities from apprenticeship to experiential learning. *Canadian Apprenticeship Journal*, 5. Retrieved from: www.adapt.it/fareapprendistato/docs/combining_workplace_training. pdf.

Schallock, B., Rybski, C., Jochem, R. and Kohl, H. 2018. Learning factory for industry 4.0 to provide future skills beyond technical training. *Procedia Manufacturing*, 23, pp. 27–32.

Skulmoski, G.J., Hartman, F.T. and Krahn, J. 2007. The Delphi method for graduate research. *Journal of Information Technology Education*, 6, pp. 1–21.

Stein, S.J. and Book, H.E. 2011. *The EQ edge: Emotional intelligence and your success*. Hoboken, NJ: John Wiley & Sons.

Stitt-Gohdes, W.L. and Crews, T.B. 2004. The Delphi technique: A research strategy for career and technical education. *Journal of Career and Technical Education*, 20(2), pp. 55–67.

Tatum, C. 2010. Construction engineering education: Need, content, learning approaches. In *Proceedings of the Construction Research Congress 2010: Innovation for Reshaping Construction Practice*, Construction Research Congress 2010, May 8–10, 2010, American Society of Civil Engineers (ASCE), Banff, AB, p. 183.

Turhan, C. and Akman, I. 2013. Employability of IT graduates from the industry's perspective: A case study in Turkey. *Asia Pacific Education Review*, 14(4), pp. 523–536. doi: 10.1007/s12564013-9278-5.

Turoff, M. 1970. The design of a policy Delphi. *Technological Forecasting and Social Change*, 2(2), pp. 140–171.

Veltri, A.T. 1986. *Expected use of management principles for safety function management* (Ph.D. dissertation, West Virginia University, Morgantown, WV).

Washer, P. 2007. Revisiting key skills: A practical framework for higher education. *Quality in Higher Education*, 13(1), pp. 57–67. doi: 10.1080/13538320701272755.

Wasonga, T.A. and Murphy, J.F. 2006. Learning from tacit knowledge: The impact of the internship. *International Journal of Educational Management*, 20(2), pp. 153–163.

Watts, A.G. 2013. Career guidance and orientation. In the United Nations Educational, Scientific and Cultural Organization (UNESCO). *Revisiting Global Trends in TVET: Reflections on Theory and Practice*, Germany. UN Campus. Retrieved from: www.unevoc. unesco.org/fileadmin/up/2013_epub_revisiting_global_trends_in_tvet_book.pdf.

Wickramasinghe, V. and Perera, L. 2010. Graduates', university lecturers' and employers' perceptions towards employability skills. *Education+ Training*, 52(3), pp. 226–244.

Woudenberg, F. 1991. An evaluation of Delphi. *Technological Forecasting and Social Change*, 40, pp. 131–150.

Zami, M.S. and Lee, A. 2009. A review of the Delphi technique: To understand the factors influencing adoption of stabilised earth construction in low cost urban housing. *The Built & Human Environment Review*, 2, pp. 37–50.

Section IV

Conclusions and recommendations

9 Conclusions and recommendations

Conclusions

Conclusion 1: generic skills

One of the objectives of the book was to ascertain the influence of generic skills (GS) in the employability framework. In achieving this, an extensive literature review was conducted on generic skills. Owing to the various dynamics and complexities facing the construction industry today such as increased global competition, technological advancements, environmental factors, and construction projects sophistication, graduates are required to possess GS if they are to attain employability. Findings from the literature review revealed several generic or non-academic skills such as communication skills, teamwork skills, analytical problem-solving skills, adaptability skills, critical thinking (creative thinking) skills, organizing and management skills, lifelong learning skills and interpersonal skills, entrepreneurship skills, leadership skills, and ethical skills. These skills were noted in the studies of Love et al. (2002), Washer (2007), Kilgour and Koslow (2009), Ahn et al. (2012), Ayarkwa et al. (2012), Reid and Anderson (2012), Jackson and Chapman (2012), Samavedham and Ragupathi (2012), Finch et al. (2013), O'Leary (2017), and Nguyen (2019). For this objective, a Delphi study was conducted and experts from the built environment identified GS as a key construct with an absolute median score of 10, highlighting their agreement on its importance on graduate employability. Findings from the Delphi study confirmed that critical thinking skills, creativity/innovation skills, willingness to learn, ICT skills, and problem-solving skills were the highest-ranked GS for graduate employability. The consensus reached by experts corroborates the findings from literature, which reinforces the need for graduates to possess these non-academic skills.

Conclusion 2: discipline-specific skills

Next was to ascertain the influence of discipline-specific skills (DSS) in the employability framework. In achieving this, an extensive literature review was conducted on the DSS which students definitely have to possess during their academic cycle. Findings from the literature review revealed that knowledge and skills acquired by studying an academic discipline include knowledge of subject area and curricula, critical thinking, pedagogical skills, organizational skills, communication skills, administrative skills, management skills, lifelong learning, decision-making skills, digital literacy, self-confidence, connection with other disciplines, problem-solving skills, collaborative skills, transformative skills, and hands-on experience. This set of skills confirms the works of Rae (2007), Bridgstock (2009), Coetzee and Beukes (2010), Hutchinson (2013), Watts (2013), Nagarajan and Edwards (2015), and Lamanauskas and Augienė (2017). For this objective, a Delphi study was conducted and experts from the built environment identified DSS as a key construct with a median score of 9, highlighting their agreement on its importance on graduate employability. Findings from the Delphi study confirmed that knowledge of the subject area and lifelong learning were the most important factors of DSS that contribute to graduate employability.

Conclusion 3: work-integrated learning

Next was to assess the work-integrated learning factors that are required for the employability of built environment graduates. Findings from the literature review revealed that the skills and knowledge obtained by undergoing WIL include developing practical knowledge, understanding job responsibilities, awareness of industry, improved work attitude, developing interpersonal values, résumé development, awareness of workplace culture, time management, self-confidence, problem-solving ability, working in multidisciplinary teams, site-supervision knowledge, interpersonal relationship with industry professionals, networking with other interns on-site, exposure to career opportunities, acquisition of professional skills, technical report writing, and mentorship opportunities. This set of skills confirms the works of Gill and Lashine (2003), Mihail (2006), Wasonga and Murphy (2006), Divine *et al.* (2007), Cooper *et al.* (2010), Sattler *et al.* (2011), Lowden *et al.* (2011), and Pillai *et al.* (2012). A Delphi study was conducted and experts from the built environment identified WIL as a key construct with a median score of 9, highlighting their agreement on its importance for graduate employability. Findings from the Delphi study confirmed that time management, developing practical knowledge, understanding job responsibilities, awareness of the industry, exposure to career opportunities, enhanced learning, and acquisition of professional skills were the most important factors of WIL that contribute to graduate employability.

Conclusion 4: emotional intelligence

This book also assessed the emotional intelligence factors that are required for the employability of built environment graduates. Findings from the literature review revealed that by possessing EI, students could benefit in the following ways: they could perceive and accurately express emotion, make decisions effectively, communicate effectively, manage relationships, show compassion for others, solve

problems flexibly, build effective relationships, adapt to new ideas and surroundings, manage stress, and display resilience in the face of adversity. This set of skills
confirms the works of Goleman (1998), Cherniss (2000), Fugate *et al.* (2004),
Lopes *et al.* (2006), Blickle and Witzki (2008), Coetzee and Beukes (2010), Stein
and Book (2011), and Jameson *et al.* (2016). From the Delphi study conducted,
experts from the built environment identified EI as a key construct with a median
score of 9, highlighting their agreement on its importance for graduate employability. Findings from the Delphi study confirmed that understanding and managing emotions, improved decision-making, flexibility in solving problems, the
ability to adapt to new ideas and surroundings, managing stress, and the ability
to display resilience in the face of adversity were the most important factors of EI
that contribute to graduate employability.

Conclusion 5: university and industry collaboration

Next was to examine the university and industry collaboration outcomes that
are required for the employability of built environment graduates. The UIC construct was one of the employability gaps introduced into this study and adequately
discussed earlier in this book. Findings from the literature review revealed that
the skills and knowledge that are achievable as a result of effective UIC include
enhancing job opportunities for graduates, encouragements of field trips to industry, industry mentoring for students, vocational training, networking activities
with industry professionals, publication opportunities with industry, patenting
opportunities, training programme opportunities for students, students' exposure
to innovative ideas, and involvement in construction projects. These roles confirm the works of Feng *et al.* (2011), Ramakrishnan and Yasin (2011), Haupt
(2012), and Turhan and Akman (2013). From the Delphi study conducted,
experts from the built environment identified UIC as a key construct with a
median score of 9, highlighting their agreement on its importance on graduate
employability. Findings from the Delphi study corroborated the findings from the
robust literature review.

Conclusion 6: 4IR skills and knowledge

Finally, the influence of 4IR skills and knowledge that are required for the employability of built environment graduates was also assessed. The 4IR construct was
one of the employability gaps introduced into this study and adequately discussed earlier in this book. Findings from the literature review revealed that the
skills and knowledge that are achievable as a result of possessing 4IR knowledge
include principles of robotics, networking and its applications, cloud-based technology, cloud computing, knowledge of cryptocurrency, photonics and quantum
materials, data analytics, drone technology, artificial intelligence, cybersecurity,
digital control systems, 3D printing application, and mixed reality. These skills
and abilities confirm the works of Madsen *et al.* (2016), Benesova *et al.* (2017),
Cotet *et al.* (2017), Grzybowska and Łupicka (2017), Pinzone *et al.* (2017), Ras

et al. (2017), Caldarola *et al.* (2018), and Schallock *et al.* (2018). From the Delphi study that was conducted, experts from the built environment identified 4IR knowledge as a key construct with an absolute median score of 10, highlighting their agreement on its importance on graduate employability. This Delphi result resonates with findings from the robust literature review.

Contributions and value of this research

Considering the present-day dynamics of the construction industry, higher education institutions around the world are under significant pressure to respond even more to the rapidly changing landscape owing to the fourth industrial revolution. One of the major contributions of this book to the body of knowledge is the development of an employability framework for built environment graduates using South Africa as a case study. The developed framework shows the influences of the exogenous constructs in predicting the overall employability of graduates, which no study has conducted in the 4IR era. One of the theories that emanated from the framework is that the employability of built environment graduates is directly related to the influence of the exogenous variables such as generic skills, discipline-specific skills, work-integrated learning, emotional intelligence, university–industry collaboration outcomes, and 4IR knowledge. The model is centred on the theory that the possession of these latent constructs will lead to graduate employability. This firmly resonates with the human capital theory which was discussed earlier in this book. The model further theorizes that employable graduates can adapt to varying demands of the industry, which results in increased productivity to themselves and their employers. Another theory developed from this model is that apart from discipline-specific skills, students require other skills and knowledge to be deemed employable. Hence, the values and contributions of this research are described under three themes: theoretical, methodological, and practical contributions.

Theoretical contributions and value

The outcomes revealed that the conceptual employability improvement framework constitutes six latent constructs (exogenous variables) which are significant to improved graduate employability (endogenous variable). Despite the debates and reviews surrounding the concept of graduate employability, there has been little mention of an employability skills framework specifically for the built environment domain in the 4IR era and conducted in South Africa, a gap which this book has addressed. The Employability Improvement Framework (EIF) that was developed has never been tested in a multidimensional structure composed of generic skills, discipline-specific skills, work-integrated learning, emotional intelligence, university–industry collaboration outcomes, and Industry 4.0 knowledge.

   Another significance of this book is that it addresses the lack of theoretical information about which of the constructs is significant in predicting the overall employability of graduates. Findings from the Delphi study conducted also indicate that the latent constructs of generic skills, discipline-specific skills, work-integrated learning, emotional intelligence, university–industry collaboration outcomes, and Industry

4.0 knowledge were found to have a significant influence in determining graduate employability. Moreover, from the literature review conducted, there was no evidence of a similar study (developing an employability skills framework for built environment graduates in the era of the fourth industrial revolution) conducted in South Africa. Most of the existing employability framework have discussed and adopted the concept in various fields such as management sciences, psychology, commerce, economics, and business studies, amongst others. With the buzz surrounding the 4IR under the administration of President Cyril Ramaphosa, this study is relevant, timely, and essential as the nation seeks to equip its students for this latest wave of digitalization. In addition, there was no evidence of an employability study (in the built environment domain) that was achieved by a two-stage Delphi process in developing a model. Therefore, this study is poised to spark an employability discussion for other researchers who may adopt this study as a base towards understanding employability in the era of the fourth industrial revolution.

Likewise, this book developed an employability skills framework using six latent constructs with the inclusion of two new constructs: university–industry collaboration outcomes and 4IR knowledge. It is worth noting that previous employability framework have used other constructs without the inclusion of the ones adopted for this study. Both of these added constructs (gaps in employability skills studies) were adequately discussed earlier in this book and were backed with several theories. More importantly, these additional constructs were validated by Delphi experts. Hence, they contribute a great deal to the employability improvement framework for this study. Therefore, this framework can be tested, generalized, and adopted as a basis for understanding the concept of graduate employability in South Africa and beyond. It is also worth noting that the CareerEDGE model provided several constructs for employability which are adopted by this book. One of the strengths of the CareerEDGE model is that it improves on other existing models such as the USEM and DOTS model (as discussed during this book). This model was developed to explain practically the concept of employability, which has been described as indefinable, particularly to students and their parents. The acronym 'CareerEDGE' was developed for easy recall, indicating that to be employable, all five elements (Career development learning, Experience – work and life, Degree subject knowledge, understanding, and skills, Generic skills, and Emotional intelligence) need to be developed. This will result in the development of self-confidence, self-efficacy, and self-esteem, which are key attributes for employability. However, the book went a step further to introduce the 4IR element into its framework owing to the current wave of digitalization. Because of the present significant and innovative changes, the concept of employability has evolved, hence the relevance of this book.

Methodological contributions and value

Many employability studies in South Africa have used various statistical methods (first-generation analysis methods) for analysis such as exploratory factor analysis (EFA), ANOVA, MANOVA, content analysis, and regression modelling to address issues relating to employability. However, this research adopted an expert-driven methodology (Delphi study) to determine the constructs that predict graduate employability and hence to develop an employability improvement framework. Owing to the advent of the fourth industrial revolution, there have been several generalized statements as to which of the latent constructs actually contribute to graduate employability. Therefore, by conducting this study using the Delphi process, it was possible to state precisely the constructs that are significant and contribute to graduate employability such as generic skills, work-integrated learning, emotional intelligence, university–industry collaboration outcomes, and 4IR knowledge. Other similar studies can validate the employability framework presented in this study. Apart from the obvious contribution to the methodological aspect of the body of knowledge, this research contributes to the employability discussions in South Africa in the wake of the advent of the fourth industrial revolution.

Practical contributions and value

Despite the exhaustive discussions on employability over the past few years, the concept remains a dynamic issue that has continued to generate diverse opinions from both industry professionals and academicians. The literature distilled in this research portrays that the employability concept is not new, dating back to the 1950s; however, its impact was more prominent in the late 1990s. Most notably, the concept has gained momentum in recent times owing to the advent of the fourth industrial revolution. This recent wave of digitalization has gathered global momentum and South Africa has continued to outline its vision to lead the 4IR march on the African continent. This innovative era represents a significant transformation as the interaction between humans and machines takes on a radical dimension.

In recent times, the South African government has embarked on several initiatives across the nation, which highlights its readiness to become a major player in the 4IR era. One of such initiatives was the inaugural Digital Economy Summit which was held at the Gallagher Convention Centre, Johannesburg, on July 5, 2019, where President Cyril Ramaphosa discussed the nation's commitment to a 'skills and technology' revolution that will boost the nation's human capital required in the 4IR era. During the summit, a live 3D hologram of the President was transmitted to the Rustenburg Civic Centre in the North West, showcasing some of the components of the 4IR elements. These emerging technologies have increased the pressure on educational institutions to introduce technology-driven courses to equip students with the required knowledge needed to thrive with the advent of the latest disruptor.

It is worth noting that there has been little research done on developing an employability skills model for built environment graduates in South Africa in the advent of the 4IR era. Most of the existing employability models have discussed and adopted the concept in various fields such as management sciences, psychology, commerce, economics, and business studies, amongst others. Hence, this book developed an employability improvement framework and identified the constructs with a direct influence (generic skills, work-integrated learning, emotional intelligence, university–industry collaboration outcomes, and 4IR knowledge) and an indirect influence (discipline-specific skills) on the employability of built environment graduates.

Furthermore, the South African educational sector has a vital role to play in realizing the nation's future goals, which is aimed at developing 'one million young people in various Industry 4.0 elements and related skills by 2030'. In achieving this milestone, the nation's commitment to improving the curriculum of higher education is certain, with various proposed initiatives positioned in place by the government. What is significant though is the nation's goal for its graduates (citizenry) to be adequately skilled to handle industry positions owing to the advent of the fourth industrial revolution. This is in line with the nation's long-term National Development Plan (NDP) which aims to develop the key capabilities and skills of its citizenry by ensuring quality education on a broader scale by 2030. This book therefore contributed to the body of employability knowledge by introducing the 4IR component into the employability discourse, something which was highlighted during the recent inaugural Digital Economy Summit held in Johannesburg. The additional knowledge derived from this study will be beneficial in achieving a more realistic understanding of the concept of graduate employability in South Africa and beyond.

South Africa's readiness for the 4IR (higher education approaches)

One of the implications of this book is that the various 4IR initiatives proposed by the South African government signal its intention of embarking on digitalization as a nation. This places significant pressures on the educational sector to revamp its curricula to accommodate several 4IR components as recommended by South Africa's President Cyril Ramaphosa. At the beginning of 2019, the President inaugurated a 30-member committee which has been tasked with leveraging opportunities presented by the fourth industrial revolution to solve contemporary issues. AI expert-scholar, Prof. Tshilidzi Marwala, Vice Chancellor of the University of Johannesburg (UJ), was appointed as the Deputy Chair of the committee. The wave of curricula restructuring has gained momentum across the African continent and South African universities are increasingly making giant strides in aligning their activities along the 4IR theme.

At the University of Johannesburg, there are existing initiatives in place to contribute to improved learning experience among students with the Global Excellence and Stature Fund (GES-Fund), one of many. As part of the university's strategy to propel the 4IR skills development initiative, the GES has begun to expand its resource base by instituting numerous flagship programmes, with the formation of the Institute for Intelligent Systems (IIS), the latest of such programmes. Launched in September 2019, the IIS is designed to further enhance the emerging concept of human–machine collaborations and augmented human intelligence. The platform will seek to bring ICT experts, intellectual researchers, and academicians together to develop innovative approaches to understand the technologies and intricacies that accompany the fourth industrial revolution. The

IIS will achieve its 4IR mandate through three main aspects: academic development, strategic research, and enterprise development. More importantly, the IIS will provide the opportunity for individuals to pursue a master's degree in both Artificial Intelligence (AI) and Financial Engineering (FE). The institute will also allow individuals to seek short learning programmes in fields such as Computational Intelligence. Similarly, the institute is expected to cover several 4IR areas such as big data analytics, predictive maintenance, systems intelligence, cyberphysical systems (CPS), optimization, cognitive computing, deep learning, machine learning, and other relevant 4IR components.

In addition, UJ has taken several steps to ensure that its students are conversant with the drastic changes expected to occur with the 4IR age and to be aware of Africa's role in this disruptive era. One of the anticipatory steps taken by the university was to appoint Africa's first and only black Nobel laureate for Literature, Prof. Wole Soyinka, in 2017. His appointment as a Distinguished Visiting Professor in the Faculty of Humanities is expected to drive the decolonization movement and further inspire Africans to develop the continent rather than develop the already advanced nations. Hence, his appointment is expected to propel Africans to use the 4IR knowledge to resolve African socio-economic problems as the continent gears up for the 4IR. From the foregoing discussion, UJ's role in Africa's acceptance and implementation of 4IR cannot be overstated. Currently, the university possesses one of the highest concentrations of PhD holders in the area of artificial intelligence on the continent. Moreover, the Vice Chancellor, Prof. Marwala, is the Deputy Chairman of the Presidential Commission on 4IR and is well versed in the AI discipline.

In recent times, the university has hosted several dialogues regarding the 4IR which include 'Higher Education 4.0, Business and Industry 4.0, Ethics 4.0, Libraries 4.0 and The Future of Work 4.0'. In fact, the university libraries have been rebranded as Library 4.0 to re-echo this new wave of revolution and to highlight the university's willingness to be part of this disruptive era. In 2019, Prof. Clinton Aigbavboa who is a Professor of Sustainable Human Settlement and Director of Sustainable Human Settlement and Construction Research Centre spearheaded a leading conference at the University with the theme 'Positioning the construction industry in the fourth industrial revolution'. The conference which was held during July 28–30 brought together industry professionals, academic researchers, government consultants, and students under one roof to deliberate on the 4IR and examine various ways to leverage the opportunities presented by this wave of digitalization.

In addition, during the early half of 2019, UJ established a 4IR partnership with New Zealand's Auckland University of Technology. Through real-time data analytics, both universities developed a World Happiness Index (WHI) for the main purpose of tracking a nation's happiness coefficient (Gross National Happiness). The scale for measuring happiness levels was from 0 to 10, with 0 indicating extreme unhappiness, 5 indicating neutral, and 10 indicating extreme happiness. Based on the results from the WHI, South Africa had a score of 4.7. Out of 156 countries indexed, South Africa was ranked 106th. Both universities have agreed to develop a real-time web portal in the future, which can be easily accessed by anyone who requires an instantaneous happiness report.

Aside from UJ, the University of Pretoria (UP) has also established several initiatives in the wake of the fourth industrial revolution. One of such initiatives is the establishment of 'Future Africa', a multidisciplinary research establishment that seeks to address complex societal problems on the continent by using science, technology, and innovation (STI). The Future Africa forum is expected to bring together ICT experts, researchers, and scientists from several disciplines to work on cutting-edge research projects with the hope of addressing major socio-economic issues on the continent. Future Africa is expected to

> focus on promoting academic leadership by developing local and African sci-ence leaders which can be achieved by increasing the number of researchers on the continent through leadership development programmes such as post-graduate, postdoctoral and research fellowships, and leverage technology to boost academic research results.

Moreover, UP launched Engineering 4.0 in August 2018, which aims to major in smart cities and transportation for the future. This initiative is aimed at har-nessing civil engineering skills, technology, and data sciences as the university fortifies its mandate to propagate the 4IR vision across the nation and beyond.

In related developments, the University of Cape Town (UCT) has established the Robotics and Agents Lab (RAL) (www.rarl.uct.ac.za/rarl/aboutus). The lab is designed to

> teach and research in the fields of Robotics, Artificial Intelligence, and Com-putational Intelligence. The main robots currently under development are in the fields of rescue and underwater exploration. In rescue, the focus has shifted away from larger and more expensive robots to smaller, cheaper and simpler ones. The underwater ROV is an inspection class vehicle designed to operate at depths of up to 300 m.

UCT is also one of the five South African universities (nodes) where the Centre for Artificial Intelligence Research (CAIR) is located. The CAIR is 'a South African distributed Research Network that conducts foundational, directed and applied research into various aspects of Artificial Intelligence'. Other South African universities where these nodes are located include North-West Univer-sity, Stellenbosch University, the University of Pretoria, and the University of KwaZulu-Natal (Gleason, 2018).

From the foregoing discussion, universities across the nation have begun to respond to the increasing pressures being placed on them due to the 4IR era. While the study highlights a few of the initiatives being undertaken by selected universities, it is worth noting that at the time of this study, several ongoing 4IR initiatives were being undertaken by several other South African universities which will be highlighted by future studies.

Recommendations

After discussing the contributions of this research, this book further makes theoretical, methodological, and practical recommendations.

Methodological recommendations

To improve the general application of this book, it is recommended that a similar employability study be conducted in both developed and developing nations with different populations and samples. Moreover, future research should be carried out to determine more latent constructs that could arise in the near future and that will contribute to graduate employability. This is because there is a possibility that graduate employability could become a function of more indicator variables. Technically speaking, there is no perfect model and as such the employability framework developed by this study can be improved upon. The study also recommends that the use of the Delphi study should be encouraged in employability studies to eliminate any weaknesses that exist by conducting traditional literature reviews.

Theoretical recommendations

Despite the exhaustive discussions on employability over the past few years, the concept remains a dynamic issue that has continued to generate diverse opinions from both industry professionals and academicians. Somehow, with the advent of the fourth industrial revolution, an employability skills framework in the built environment domain is yet to be developed, a gap which this study had filled. This current book reviewed existing literature and conducted a robust review of nine existing employability models across several contexts. From the nine employability models, this study identified four constructs that contribute to employability. The study also considered the inclusion of two constructs (research gaps) which resulted in six latent constructs. These were validated by built environment experts during a two-stage Delphi process which led to an employability improvement framework. The constructs identified were generic skills, discipline-specific skills, work-integrated learning, emotional intelligence, university–industry collaboration outcomes, and Industry 4.0 knowledge. It is therefore recommended that the employability framework developed in this study should form the basis for understanding the concept in South Africa and beyond in the wake of the 4IR era.

Policy implication and practical recommendations

Based on several employability discussions in which this book has engaged, the following policy implications and practical recommendations have been identified:

1   As discussed throughout this research, the role of higher education in developing employability skills cannot be overstated. Hence, the

implication of this book for higher education is to respond to the pressures it faces by enforcing meaningful and relevant teaching activities to ensure their curricula develop graduates who are well equipped to handle industry positions in both design and supervisory roles after graduation. Institutions of higher learning also need to be more innovative in designing their pedagogical approaches as they demonstrate their reliability in producing quality learning outcomes for students. Apart from being academically sound, the industry requires graduates who are sufficiently furnished with the relevant skills, abilities, and competencies to fit into the world of work after graduation.

2   It is recommended that higher education curricula sufficiently strike a balance between theoretical knowledge and practicality to provide a holistic approach in developing employability skills among students. Most of the pedagogical approaches discussed in this book are underpinned by the concept of active learning approaches in which students are actively engaged in the learning process. Unlike passive approaches such as traditional lectures, active-based learning promotes critical thinking skills, increases students' engagement, increases retention, and fosters problem-solving initiatives.

3   Owing to the advent of the fourth industrial revolution, higher education will be required to be more innovative when designing pedagogical approaches in developing employability skills. Therefore, future research studies are to be conducted to expound more on these innovative teaching methods in this era of digitalization. Also, a follow-up research should be conducted to predict the various employability skills that employers will seek in the near future owing to the advent of the fourth industrial revolution. This is because the skills are changing in real time.

4   The fourth industrial revolution is not a concept for the future; it is upon us! As automation continues to advance rapidly, the pressures on universities to revamp and restructure their curricula (skills revolution) have become increasingly necessary. This can be achieved by introducing significant modifications to the science and technology curricula to enable students to develop competencies in the rapidly emerging areas of artificial intelligence, data science, robotics, advanced simulation, data communication, system automation, real-time inventory operations, cybersecurity, and information technologies. This implies that universities' curricula should inculcate 4IR components within the conventional primary sciences such as biology, chemistry, and physics, with greater emphasis on digital literacy to boost 4IR understanding among young learners. As discussed earlier, several universities across the nation have already begun 4IR initiatives to develop digital-savvy students. Also as discussed earlier, the 4IR era combines both physical and virtual reality, hence students are to be immersed in virtual learning environments to fully grasp the concept of the innovativeness surrounding the fourth industrial revolution. As previously indicated, universities around the nation have begun the integration of 4IR components into

their curricula which bodes well for the future workforce and aligns with President Ramaphosa's 4IR mandate.

5   Finally, with the advent of the fourth industrial revolution, conducting this study across other developing countries could be considered.

Limitations

Despite the valuable insights provided by this book around the concept of graduate employability in South Africa, there are some limitations surrounding the study. This research was conducted in the Gauteng Province of South Africa. This is because although it is the smallest province in South Africa based on land area, the city nicknamed 'Place of Gold' has a high population density, making it the most populated province in the country with approximately 14.7 million people. Gauteng was selected because the city contributes enormously to various sectors in the country such as construction, manufacturing, technology, and finance, amongst others, making it the economic hub of Mzansi. Hence, Gauteng is a typical representation of the entire urban space in South Africa and was adopted as the study population. With enough time and resources, a similar study can be conducted across the other eight provinces across South Africa. Also, there are several possible areas of employability that need to be considered. This study did not elaborate more on other grey areas of employability, which is another limitation of this study.

Finally, this study developed an employability framework based on six latent constructs: generic skills, discipline-specific skills, work-integrated learning, emotional intelligence, university–industry collaboration outcomes, and Industry 4.0 knowledge. It is acknowledged that this framework can be further refined to include constructs such as external factors and personal circumstances, culture, and other relevant constructs, which were not considered for this study.

Recommendations for future research

Based on some of the limitations observed, the following suggestions were made for further studies:

1   Further employability studies should be conducted across South Africa and around the continent to test the employability measurement scales (constructs) rigorously to further the knowledge of the employability discussions. The measurement scales and exogenous constructs used for this study may not apply to other contexts or professionals; hence the need for this study's results to be replicated with a different set sample frame and population.

2   This study (model) did not consider the influence of culture (as a distinct construct) on graduate employability. It is therefore recommended that future studies should examine this aspect as culture embraces several factors such as values, attitudes, and beliefs, which can directly or indirectly contribute to employability outcomes.

Conclusions

This study developed an Employability Improvement Framework by adopting latent constructs from existing employability models and theories. The study postulated that graduate employability is directly related to the influence of six exogenous variables in predicting the outcomes of improved graduate employability. Therefore, these latent constructs schematically represented adequately address graduate employability in South Africa in the era of the fourth industrial revolution. This study obtained responses from built environment professionals during the Delphi study, hence the results of this study have theoretical, methodological, and practical implications. The knowledge derived from this study is beneficial to achieving a more realistic understanding of the employability concept. For individuals who are involved in skilled human resource development, this study is valuable because it enlightens policymakers on employability skills that graduates require to fit easily into the world of work.

The findings emanated from this book reinforce the need to establish effective collaborations between higher education and the construction industry to produce competent graduates who can effectively thrive after graduation. Moreover, the employability framework developed in this book will provide a solid reference for researchers who seek to explore the employability concept beyond the scope of this study. While other employability studies across South Africa have adopted alternative statistical methods to discuss some of the constructs, this study lends its support to the body of knowledge by employing an innovative Delphi approach. It is worth noting that presently, an academic degree does not guarantee employability but the possession of both academic and non-academic skills, which were adequately discussed throughout this book.

References

Ahn, Y.H., Annie, R.P. and Kwon, H. 2012. Key competencies for US construction graduates: Industry perspective. *Journal of Professional Issues in Engineering Education and Practice, 138*(2), pp. 123–130. doi: 10.1061/(ASCE)EI.1943-55410000089.

Archer, W. and Davison, J. 2008. *Graduate employability: What employers want?* London: The Council for Industry and Higher Education.

Ayarkwa, J., Adinyira, E. and Osei-Asibey, D. 2012. Industrial training of construction students: Perceptions of training organizations in Ghana. *Education Training, 54*(2/3), pp. 234–249. http://dx.doi.org/10.1108/00400911211210323.

Benesova, A. and Tupa, J. 2017. Requirements for education and qualification of people in Industry 4.0. *Procedia Manufacturing, 11*, pp. 2195–2202.

Blickle, G. and Witzki, A. 2008. New psychological contracts in the world of work: Economic citizens or victims of the market? *Society and Business Review, 3*(2), pp. 149–161.

Bridgstock, R. 2009. The graduate attributes we've overlooked: Enhancing graduate employability through career management skills. *Higher Education Research & Development, 28*(1), pp. 31–44. doi: 10.1080/07294360802444347.

Caldarola, E.G., Modoni, G.E. and Sacco, M. 2018. A knowledge-based approach to enhance the workforce skills and competences within the industry 4.0. *Proceedings of the*

Tenth International Conference on Information, Process, and Knowledge Management, Rome.

Cherniss, C. 2000. Emotional intelligence: What it is and why it matters. *Paper presented at the annual meeting of the Society for Industrial and Organizational Psychology*, New Orleans, USA.

Coetzee, M. and Beukes, C.J. 2010. Employability, emotional intelligence and career preparation support satisfaction among adolescents in the school-to-work transition phase. *Journal of Psychology in Africa, 20*(3), pp. 439–446.

Cooper, L., Orrell, J. and Bowden, M. 2010. *Work integrated learning: A guide to effective practice*. New York: Routledge.

Cotet, G.B., Balgiu, B.A. and Zaleschi, V.C. 2017. Assessment procedure for the soft skills requested by industry 4.0. *MATEC Web of Conferences*, 121, p. 07005. EDP Sciences.

Divine, R.L., Linrud, J.K., Miller, R.H. and Wilson, J.H. 2007. Required internship programs in marketing: Benefits, challenges and determinants of fit. *Marketing Education Review, 17*(2), pp. 45–52. http://dx.doi.org/10.1080/10528008.2007.11489003.

Feng, C., Ding, M. and Sun, B. 2011. A comparison research on industry-university-research strategic alliances in countries. *Asian Social Science, 7*(1), p. 102.

Finch, D.J., Hamilton, L.K., Baldwin, R. and Zehner, M. 2013. An exploratory study of factors affecting undergraduate employability. *Education Training, 55*(7), pp. 681–704. http://dx.doi.org/10.1108/ET-07-2012-0077.

Fugate, M. and Ashforth, B. 2004. Employability: The construct, its dimensions, and applications. *Proceedings of the Annual Meeting of the Academy of Management, OB: J1-J6*, Seattle.

Gill, A. and Lashine, S. 2003. Business education: A strategic market-oriented focus. *International Journal of Educational Management, 175*, pp. 188–194. http://dx.doi.org/10.1108/09513540310484904.

Gleason, N.W. (Ed.). 2018. *Higher education in the era of the fourth industrial revolution*. Singapore: Palgrave Macmillan, Springer Nature.

Goleman, D. 1998. *Working with emotional intelligence*. New York: Bantam.

Grzybowska, K. and Łupicka, A. 2017. Key competencies for industry 4.0. *Economics & Management Innovations, 1*(1), pp. 250–253.

Haupt, T. 2012. Student attitudes towards cooperative construction education experiences. *Construction Economics and Building, 3*(1), pp. 31–42.

Hutchinson, J. 2013. *School organisation and STEM career-related learning*. Derby: International Centre for Guidance Studies. Retrieved from: http://derby.openrepository.com/derby/bitstream/10545/303288/8/STEM%20Leaders%20Report%202013%20%28High%20res%29.pdf.

Jackson, D. and Chapman, E. 2012. Non-technical competencies in undergraduate business degree programs: Australian and UK perspectives. *Studies in Higher Education, 37*(5), pp. 541–567. doi: 10.1080/03075079.2010.527935.

Jameson, A., Carthy, A., McGuinness, C. and McSweeney, F. 2016. Emotional intelligence and graduates – employers' perspectives. *Procedia – Social and Behavioural Sciences, 228*, pp. 515–522.

Kilgour, M. and Koslow, S. 2009. Why and how do creative thinking techniques work? Trading off originality and appropriateness to make more creative advertising. *Journal of the Academy of Marketing Science, 37*(3), pp. 298–309.

Lamanauskas, V. and Augienė, D. 2017. Career education in comprehensive schools of Lithuania: Career professionals' position. In A. Klim-Klimaszewska, M. Podhajecka and

A. Fijalkowska-Mroczek (Eds.), *Orientacje i przedsiewziecia w edukacjiprzedszkolnej i szkolnej /Monografia/* (pp. 216–236). Siedlce: Akka.

Lopes, P.N., Grewal, D., Kadis, J., Gall, M. and Salovey, P. 2006. Evidence that emotional intelligence is related to job performance and affect and attitudes at work. *Psichothemia*, 18(1), pp. 132–138.

Love, P., Haynes, N., Sohal, A., Chan, A. and Tam, C. 2002. *Key construction management skills for future success*. Monash University. Faculty of Business and Economics. Working Paper 49/02.

Lowden, K., Hall, S., Ellio, D.D. and Lewin, J. 2011. *Employers' perceptions of the employability skills of new graduates*. London: Edge Foundation.

Madsen, E.S., Bilberg, A. and Hansen, D.G. 2016. Industry 4.0 and digitalization call for vocational skills, applied industrial engineering, and less for pure academics. *Proceedings of the 5th P&OM World Conference, Production and Operations Management, P&OM*, Havana International Conference Center, Havana.

Mihail, D.M. 2006. Internships at Greek universities: An exploratory study. *Journal of Workplace Learning*, 18(1), pp. 28–41. doi: 10.1108/13665620610641292.

Nagarajan, S. and Edwards, J. 2015. The role of universities, employers, graduates and professional associations in the development of professional skills of new graduates. *Journal of Perspectives in Applied Academic Practice*, 3(2).

Nguyen, P. 2019. *Enhancing the employability of graduate students with transversal skills* (Bachelor's thesis, Lahti University of Applied Sciences, Lahti).

O'Leary, S. 2017. Graduates' experiences of, and attitudes towards, the inclusion of employability-related support in undergraduate degree programmes; trends and variations by subject discipline and gender. *Journal of Education and Work*, 30(1), pp. 84–105.

Pillai, S., Khan, M.H., Ibrahim, I.S. and Raphael, S. 2012. Enhancing employability through industrial training in the Malaysian context. *Higher Education*, 63(2), pp. 187–204. doi: 10.1007/s10734-011-9430-2.

Pinzone, M., Fantini, P., Perini, S., Garavaglia, S., Taisch, M. and Miragliotta, G. 2017. September. Jobs and skills in Industry 4.0: An exploratory research. *Proceedings of the IFIP International Conference on Advances in Production Management Systems*, Springer, Cham, pp. 282–288.

Rae, D. 2007. Connecting enterprise and graduate employability: Challenges to the higher education culture and curriculum? *Education+ Training*, 49(8/9), pp. 605–619.

Ramakrishnan, K. and Yasin, N.M. 2011. Higher learning institution-industry collaboration: A necessity to improve teaching and learning process. *Proceedings of the Computer Science & Education (ICCSE) 2011 6th International Conference on IEEE*, Singapore, p. 1445.

Ras, E., Wild, F., Stahl, C. and Baudet, A. 2017, June. Bridging the skills gap of workers in industry 4.0 by human performance augmentation tools: Challenges and roadmap. *Proceedings of the 10th International Conference on Pervasive Technologies Related to Assistive Environments*, Association for Computing Machinery, New York, NY, pp. 428–432.

Reid, J.R. and Anderson, P.R. 2012. Critical thinking in the business classroom. *Journal of Education for Business*, 87(1), pp. 52–59.

Samavedham, L. and Ragupathi, K. 2012. Facilitating 21st century skills in engineering students. *Journal of Engineering Education Transformations*, 25(4–1), pp. 37–49.

Sattler, P., Wiggers, R.D. and Arnold, C. 2011. Combining workplace training with post-secondary education: The spectrum of work-integrated earning (WIL) opportunities from apprenticeship to experiential learning. *Canadian Apprenticeship Journal*, 5.

Retrieved from: www.adapt.it/fareapprendistato/docs/combining_workplace_training. pdf.

Schallock, B., Rybski, C., Jochem, R. and Kohl, H. 2018. Learning factory for industry 4.0 to provide future skills beyond technical training. *Procedia Manufacturing, 23*, pp. 27–32.

Stein, S.J. and Book, H.E. 2011. *The EQ edge: Emotional intelligence and your success.* Hoboken, NJ: John Wiley & Sons.

Turhan, C. and Akman, I. 2013. Employability of IT graduates from the industry's perspective: A case study in Turkey. *Asia Pacific Education Review, 14*(4), pp. 523–536. doi: 10.1007/s12564013-9278-5.

Washer, P. 2007. Revisiting key skills: A practical framework for higher education. *Quality in Higher Education, 13*(1), pp. 57–67. doi: 10.1080/13538320701272755.

Wasonga, T.A. and Murphy, J.F. 2006. Learning from tacit knowledge: The impact of the internship. *International Journal of Educational Management, 20*(2), pp. 153–163.

Watts, A.G. 2013. Career guidance and orientation. In the United Nations Educational, Scientific and Cultural Organization (UNESCO). *Revisiting Global Trends in TVET: Reflections on Theory and Practice.* Germany: UN Campus. Retrieved from: www.unevoc. unesco.org/fileadmin/up/2013_epub_revisiting_global_trends_in_tvet_book.pdf.

Index

Note: Page numbers in *italics* indicate a figure and page numbers in **bold** indicate a table on the corresponding page.

For Product Safety Concerns and Information please contact our EU
representative GPSR@taylorandfrancis.com
Taylor & Francis Verlag GmbH, Kaufingerstraße 24, 80331 München, Germany